Your Mitoc

Key to Health and Longevity

By Warren L. Cargal, L. Ac.

Second Edition 2018

Paperback ISBN: 978-1-950282-55-5

eBook ISBN: 978-1-950282-56-2

DISCLAIMER

Neither the author nor the publisher is engaged in rendering professional advice or services to the individual reader. The information in this book is for educational purposes only and isn't meant to diagnose or in any way replace qualified medical supervision. The author is at no time prescribing any medical recommendations or diagnosing any health condition. Any suggestions that are being made are in accordance to the author's training and experiences in his field of expertise. For any medical condition, each individual is recommended to consult with his or her health care provider before using any information, idea, or products discussed herein. Neither author nor publisher shall be liable or responsible for any loss or damage allegedly arising from any information or suggestion in this book. While every effort has been made to ensure the accuracy of the information presented, neither the author nor publisher assumes any responsibility for errors. The information provided about supplements has not been evaluated or approved by the Food and Drug Administration. The supplements discussed in this book are not intended to treat, diagnose, prevent, or cure any condition or disease.

CONTENTS

ACKNOWLEDGMENTS

For me, writing this book wouldn't have been possible without the internet and some of the great tools now available to make this process easier. For those of you considering writing a book, I thought it may be helpful to share with you my approaches to research and writing. Hopefully, this will make your journey easier. And if someone has a more organized approach, please share it with me!

First, there's a great deal of quality research now available on the web. Many full articles are available for free through open access, and others you may need to purchase. Managing your source material and citations can take a lot of time if you don't organize this process effectively. Here are the tools I found most helpful:

Zotero.com, which is a free tool that allows you to collect all of your research in a single, searchable interface. There are browser extensions that allow you to save articles directly into the Zotero interface while conducting internet searches.

Readcube.com, which is available for free, or you can upgrade to the professional version that offers more editing functionality. The issue this tool resolves very nicely is being able to read an article, highlight content you want to save for a quote, and have that content saved with a corresponding reference.

Kindle reader, which is where all my book reading is done. Kindle reader allows content to be highlighted, notes to be added, and content to be exported with references.

Without a doubt, you wouldn't have this book in your hands today if it weren't for the diligent work of a couple of other people, too. Mary Allison Geibel, who was working on her

Masters of Public Health with a concentration in cancer epidemiology from Emory University while doing research for me, deserves credit for much of the cardiovascular and cardiometabolic chapters.

John S. Colman is another person without whom this book couldn't have been possible. John was formerly Product Development & Research Director at Life Extension Foundation for twenty-one years. John is a world-renowned authority on antioxidants, mitochondrial health, and aging. He specializes in free-radical chemistry and biological systems, with special emphasis on the molecular and cellular biology of aging.

John has been responsible for the development and launch of launch of twenty-seven major, groundbreaking nutraceuticals—totaling more than sixty supplements, which, thanks to his leading-edge, combined expertise in the science of aging, nutrition, and nutritional ingredients have resulted in very tangible health benefits to consumers worldwide.

John painstakingly went through my work and added his technical knowledge of the molecular biology of aging and free-radical chemistry. Additionally, he refined the nutraceutical protocols suggested in the book by adding information from the latest research.

I also want to acknowledge the work of the husband-and-wife team of Dr. James Morré, who passed away in 2016, and Dorothy Morré, PhD, and their five decades of research they've done, cumulating in the scientific book *ECTO-NOX Proteins: Growth, Cancer and Aging* (Springer, 2013). This book offers a well-organized record of years of research into aging and cancer by the Morrés, which they mainly conducted as professors at Purdue University in West Lafayette, Indiana. The late Dr. Morré, the Dow Professor of Medicinal Chemistry, was also the founding director of Purdue's cancer research center.

Their work has been instrumental in helping me to formulate many of the mechanisms involved in aging and age-related diseases, along with helping to identify the potential for nutraceutical applications.

I hope this book will enjoy a wide readership; its focus is pertinent to our health as individuals and as a society dealing with ever-increasing health care costs. For readers who possess expertise in these areas, I'd like to make clear that I haven't been able to cover every aspect of the relevant science, and you may identify aspects that may seem deficient. It's been a challenge to take the technical science and present it to a general audience. As a result, I did have to make choices on excluding some molecular pathways in order to tell the story.

Finally, the book wouldn't have been possible without the unwavering intellectual and emotional support of my beloved wife, Kimberly Cahill.

Warren Cargal

INTRODUCTION

This is a book about extending your lifespan—and, within this extended lifespan, living a healthy, disease-free life. For a number of years, there has been research on how to do this, and new research continues to confirm the importance of our mitochondria in aging and age-related diseases.

Over the last century, there have been several documented cases of humans living past 120 years, meaning that life after 50 could be the bulk of your lifespan, and attention to longevity may be the most important decision you make concerning your health and quality of life. At the core of this longevity are lifestyle and dietary choices as well as various types of health maintenance and preventive nutraceuticals and herbs that have emerged in recent decades. The real key is how we can use all of these tools to improve our lives.

Individuals concerned about healthy aging should understand that aging isn't fundamentally about superficial appearance but rather that appearance mirrors their underlying cellular maintenance as they age. As each of us looks at our own health maintenance as we get older, we must make choices concerning how best to promote healthy skin, hair, muscle, joints, organs, and—most importantly—our brain.

Individuals who use these approaches will look good for their age—youthful and not artificially young—and their health and vitality will be evident to all around them. The first step in adopting an individualized protocol of anti-aging—or, rather, healthy longevity—is to become educated about the science of our bodies as they age, which is the intent of this book.

In actuality, this science has been around for a very long time, and current research is essentially confirming the wisdom of the science of traditional Chinese medicine. Ancient Chinese medical texts describe this practice of longevity in Chinese medicine as yang sheng, or nurturing life.

Lest you get disappointed and think this is a book about Chinese medicine, rest assured it is not. This book is based on the latest, cutting-edge research, but it's surprising how the latest research validates the rich legacy of ancient Chinese medicine.

Let me provide some background on how this book came about. As you may guess, my training is in traditional Chinese medicine, which includes acupuncture and Chinese herbal formulas and lifestyle practices. For me, Chinese medicine is an observational medicine that closely matches my own life, where there have been a series of "that's interesting" observations that culminated in many "aha" moments. The following is a roadmap of some of those valuable aha moments, which ended with the realization that I needed to write this book.

Traditional Chinese medicine, or Daoist medicine, is based on theories that our cells are born with a quality that ancient Daoists called jing, commonly translated as "essence." The jing is passed from the parents to the fetus at conception—in particular, from the kidneys of the mother. Jing determines basic constitution, strength, and vitality, and the preservation of jing is integral to the ideas of longevity. Many Chinese herbal medical practices are devised to preserve the jing.

At some point, I heard the term "Mitochondrial Eve." We'll discuss mitochondria throughout this book, but, briefly, mitochondria are rod-shaped organelles (a specialized component of a cell that has a specific function) that are the power generators of the cell, converting oxygen and nutrients into adenosine triphosphate, the energy that powers the cell's metabolic activities.

Mitochondria are inherited maternally, so if we trace our genetic lineage from child to mother to maternal grandmother and so on, Mitochondrial Eve would be the mother of all mothers. She's thought to have lived in Africa approximately 170,000 years ago. This doesn't necessarily mean she was the first human or even the first female; it only means she's the most recent ancestor common to all humans living today. The reason we can trace our ancestry this way is because all mitochondria have their own DNA (genes), which is normally passed on to our children only in the mother's egg, not in the father's sperm.

That was one of those "that's interesting" observations. Ancient Chinese medicine had theorized that our basic essence, jing in the Chinese medicine world, was being passed down from the mother. And current research indicates that there exists a Mitochondrial Eve.

Another ongoing "that's interesting" observation relates to the current research on Chinese herbs. Traditional Chinese medicine has an extensive history—going back several thousand years—of using tonics made from herbs to support health and extend lifespan as well as using Jing (essence) formulas to extend one's life.

Not surprisingly, research into the physiological actions of herbal tonics has exploded in the last twenty years. Two jing "superstar" herbs that have received a lot of attention lately and are discussed in this book are astragalus (huang qi, and Polygonum cuspidatum (Hu Zhang).

Astragalus has been used in traditional Chinese medicine for thousands of years, often in combination with other herbs, to tone and strengthen the body. Modern research shows that astragalus slows formation of advanced glycation end products (AGEs) and supports learning and memory—both of which are discussed in later chapters.

Astragalus is also the primary active ingredient in traditional Chinese medicine formulas for cardiovascular support. A recent study supporting this benefit found astragalus to be especially effective at protecting the aorta from damaging effects of high levels of free fatty acids (FFAs).

P. cuspidatum historically was used in Chinese medicine for quickening blood circulation to dissipate stasis, dispelling wind, removing obstructions in the acupuncture channels, clearing heat, and promoting diuresis and detoxification. Recent research has identified that P. cuspidatum contains an isolate called resveratrol, which activates an enzymatic chemical called sirtuin 1. This compound is required for the cellular autophagic response (clearing and replacing of cellular components). Research also shows that caloric deprivation, or patterns of fasting, plays a significant role in this autophagic response that appears at the heart of longevity mechanisms. This research suggests that all of these components play a symbiotic role in a holistic treatment protocol for longevity.

Compare the recent research to the methodical development and refinement of Chinese medicinal tonic herbs and formulas that occurred over centuries—and, in some cases, thousands of years. This slower, empirical approach resulted in a huge and sophisticated body of knowledge, gathered over the course of dozens of generations and lifespans, that revealed (with a high degree of accuracy) which herbs and herbal formulas work best, which ones are problematic, and which ones belong to the class of jing tonics recognized for their ability to safely support "radiant health," defined in traditional Chinese medicine as "health beyond danger."

Again, this was one of those "that's interesting" observations. It was with a sense of wonderment that I discovered how the depth of holistic knowledge passed down over the centuries aligns with what current research confirms.

My observations began to coalesce around my clinic practice. I work with patients who are dealing with fertility issues, using acupuncture, Chinese herbs, and lifestyle changes with good success. Occasionally, I'd have a patient who did well on the protocol and would get pregnant, but, within the first three months, the pregnancy would fail.

As I researched this problem, several factors stood out that could be causative.

First, during the first three months of life, the embryo goes through a tremendous growth spurt, going from a single cell to millions of specialized cells. This growth requires energy.

Second, a girl has about one million undeveloped eggs in her ovaries at birth, but only when she reaches puberty and begins ovulation that select, immature eggs begin to move through a one-year cycle, culminating in a mature egg ready for ovulation. During the development phase, the egg is under the influence of the body's nutritional status—and requires energy to develop.

The fully developed egg, when released from the ovary, contains about one hundred thousand mitochondria. Now, compare that to a single sperm, which contains approximately one hundred mitochondria. To visualize the relationship between the egg and the sperm, imagine a large planet (the egg) circled by countless little rocket ships (sperm) looking for entry.

The mitochondria contained within the egg supply the energy to accomplish this first trimester of rapid growth. If the mitochondria are defective and unable to provide the necessary energy to sustain this rapid growth, there will be a miscarriage. With that understanding, I modified the fertility treatment protocols to avoid sugars and carbohydrates, and I incorporated supplements that specifically targeted the mitochondria energy-production pathways.

At that time, I looked at the mitochondria as a problem related to egg quality—yet another "that's interesting" observation—and just moved on in my practice. It wasn't until a couple of years ago that the topic of mitochondria came up again.

I was working with a patient who'd recently suffered a stroke and was experiencing symptoms of post-stroke sequelae (aftereffects). From a Chinese medicine perspective, the earlier the symptoms are treated after the stroke, the greater the chance for improvement.

To that end, I was researching Chinese herbs or isolates of herbs to help nourish the heart and improve oxygen supply, and I came across a book entitled The Sinatra Solution. It was written by Stephen Sinatra, MD, a cardiologist. The book is an easy weekend read and really not very technical, but is well referenced.

The book was profound for me. First, it pinpointed dysfunctional mitochondria as driving most cardiovascular diseases The dysfunction does not refer to genetics but rather lifestyle issues, such as excessive consumption of carbohydrates and sugar, lack of exercise, and shallow breathing—all of which hinder the mitochondria's ability to produce energy for a constantly beating heart.

Secondly—and this one was amazing for me—a cardiologist (Sinatra) who would normally suggest bypass surgery or heart stents was recommending nutritional protocols for cardiovascular issues. Most importantly, many of the nutritional support strategies Sinatra recommended were those I was already using in the fertility protocols to strengthen egg quality. This was exciting for me, as I realized this mitochondrial approach extended well beyond heart and fertility issues and, in fact, represented a holistic approach to our health.

This was my big "aha" moment and the inspiration to write this book. In essence, this is a continuation of a long history in Chinese medicine of writing about life-extending formulas and practices.

Fundamentally, there's a root cause addressed in this book—a root cause of increased body fat and reduced lean muscle mass, low energy levels and inefficient metabolism, increased low-grade inflammation, inadequate performance, accelerated aging, and, unfortunately for some, premature death. The treatment principle is stated in Chinese medicine as follows:

Yi bing tong zhi;	One disease, different treatments;
Tong bing yi zhi.	Different diseases, one treatment.

This statement means that patients with the same disease diagnosis may receive entirely different treatments if their presenting patterns are different. Conversely, patients with different disease diagnoses may receive essentially the same treatments if their presenting patterns are the same.

While disease aspects of our general health might seem vastly diverse and impossibly related to a single cause, you'll learn, while reading this book, there is a single causative factor—our mitochondria, which can cause one disease pattern or many different disease patterns. I can assure you, of all the important components of the cells in our bodies, none are more important than mitochondria. However, this book will show you how to take control of and increase your energy reserves to make them more efficient. The far-reaching benefits will impact every aspect of your general health and well-being.

STRUCTURE OF THE BOOK

I've tried to write this book from the position that, for many of us, biology wasn't our favorite subject. I've simplified some of the technical concepts and minimized the use of the nomenclature specific to the cellular concepts. That being said, there are some technical terms involved in explaining the functioning of the mitochondria. I encourage you to do the best you can with terms and explanations that aren't familiar to you. All of the foundational concepts flow through the chapters of the book, so you'll gain understanding as you keep reading.

Additionally, I've included links to instructional podcasts to explain some of the mitochondria concepts. If you find yourself getting stuck, try listening to the podcast. And if you find yourself getting frustrated, keep in mind your purpose for reading the book: a healthier and longer life.

Finally, if you have a technical background or are a health practitioner, I'm including a summary below of each of the chapters if you decide to skip around.

Chapter 1 provides a basic discussion of the role mitochondria play in our body and their biological origin.

Chapter 2 discusses the molecular machinery of our bodies, using a metaphor of the cell as a city and the nucleus as the city library. The highways are microtubes, and the movers of material down this highway are the kinesins. We'll walk down this highway, where each step requires an ATP (adenosine triphosphate) molecule. We'll describe how mitochondria can be pulled to different locations.

The cellular storm of atoms and molecules is dominated by

continuous motion, which scientists call thermal motion. At room temperature, atoms and molecules can travel at speeds exceeding that of the fastest jet.

We'll bring into focus the magnitude of what takes place in the cells and the mitochondria. What the reader sees in print are one-dimensional pictures, and this gives a distorted notion that events in the body are fixed and rigid. We'll show that what transpire in cells, mitochondria, and the body as a whole are very dynamic, rapidly moving three-dimensional events.

Chapter 3 describes the detailed process of mitochondrial energy generation. Furthermore, glycolysis, an alternative cellular-energy pathway, is explained in detail, as well as how it differs from aerobic respiration.

The mitochondrial electron transport chain is also explained in detail. The Krebs cycle, commonly referred to as the citric acid cycle, is described in full as well. ATP synthesis, including the molecular machinery responsible for its production, is delineated.

Chapter 4 describes the most advanced theory of aging, combining the work of Aubrey de Grey with mitochondrial mutations as well as clonal expansion of runaway mitochondria. This includes a discussion of how these mitochondria travel through the body to trigger faraway oxidative events, including cardiovascular disease. An advanced theory of membrane damage and an explanation of the plasma membrane redox system (PMRS)—a group of reductases that serve to pass electrons into and across the plasma membrane—is also described.

Morré's PMET (plasma membrane electron transport) membrane research grew out of a failure to explain longevity through free radicals and oxidative stress. Longevity is often associated with resistance to oxidative injury within and among species, but most attempts to slow down aging by antioxidant treatments have failed to show beneficial effects, prompting a new explanation for aging that involved cellular membranes.

Chapter 5, the cardiovascular chapter, takes a deeper look into different heart conditions and the role of mitochondrial dysfunction in these disorders. The specific effects of low oxygen on the heart will be addressed along with consideration of the effects of D-ribose, coenzyme Q10, L-carnitine, and certain medications for mitochondrial health of the cardiac system. This protocol is believed to correct oxygen imbalance in the heart and restore mitochondrial function.

The effects of exercise on cardiac mitochondrial function will be covered in detail, including a basic protocol to improve mitochondrial content and function, specifically in relation to the heart.

Chapter 6 discusses cardiometabolic syndrome, focusing on the prevalence of diabetes and diabetes-associated health issues. The role of mitochondria is discussed in terms of their central role in insulin secretion. The fact that mitochondrial levels decline in the liver and pancreas and in skeletal muscles in patients with type 2 diabetes is examined. Strategies focus on increasing mitochondrial function with a basic protocol specific to diabetes.

Chapter 7 discusses neurodegeneration and the role that mitochondria play in brain-specific degenerative processes. The importance of mitochondria in nerve-impulse transmission is detailed. A concise discussion of the effects of mitochondrial dysfunction centers on oxygen-based and nitrogen-based free-radical production from defective mitochondria. These byproducts can cause cell-membrane lipid peroxidation, pleating of protein sheets, beta-amyloid formation, neurofibrillary entanglement etiology, and tau formation. A basic protocol to address this damage provides specific advice on brain nerve health in relation to nerve-impulse transmission. This protocol includes neurite and dendrite health—the telecommunications network in the brain that's capable of regeneration.

Chapter 8 reviews the prevalence of arthritic conditions, focusing on how mitochondria affect joint chondrocytes. Arthritis and reduced oxygen content as a result of cardiovascular issues are covered. A basic arthritis protocol is outlined specific to anti-inflammatory support and management.

Chapter 9 describes how mitochondrial function drives skin health and how advanced glycation end products—or AGEs—contribute to skin aging. This chapter outlines a basic protocol for healthy skin support, with new additions and insights offered.

Chapter 10 describes, in detail, the hallmark of cancer—rapid abnormal cell growth—and how this occurs. Mitochondrial dysfunction is viewed as a driver of oncogenesis (development of cancerous cells). The Otto Warburg theory of cancer is described in a modified form, with emphasis on our improved understanding of the glycolytic process and oncogenesis. Herbal protocols—involving closing membrane transporters to stop tumor cell division and to help drive or restore apoptosis—are discussed in detail. The effects of calorie restriction coupled with a modified ketogenic diet are outlined.

Appendix 1 provides a summary of all the nutritional protocols discussed, which includes the Nutritional Foundation Protocol and a detailed discussion of its effects on mitochondria with special reference to coenzyme Q10, L-carnitine, D-ribose, magnesium citrate, PQQ (pyrroloquinoline quinone), resveratrol, and shilajit.

Appendix 2 describes the role of calories in mitochondrial health.

Some comments about the book before you dig in. It's based on current molecular and cellular research, and, as such, some parts may be a bit technical in nature—especially if you don't have a science or biology background. To effectively communicate the importance of the mitochondria and the significance of the research contained in this book, I have to discuss a few technical details and ensure that all readers have at least a basic understanding of cell biology. Therefore, I feel a quick-and-dirty review is well worth the few extra pages of reading. If I lose you with the details, don't get tied up in a knot; just try to understand the bigger picture. I've tried to offset this technical detail issue by including links to YouTube videos If it gets too detailed, click on one of the YouTube links for a more visual explanation or demonstration. They're great, and, as I said in the Acknowledgments section, it's amazing to see all the great tools available on the internet for learning and accessing information.

I hope you enjoy the read. Here's to your health, happiness, and longevity.

CHAPTER 1

Your Mitochondria

"Your genetics load the gun. Your lifestyle pulls the trigger."
—Mehmet Oz

At some point in your life, you realize you're aging. This realization doesn't pertain so much to your biological age—as today it's not unusual to see people in their sixties and seventies70s who appear to be in their fifties—but aging in the sense of how you feel and function. Here are some signs you might be aging:

- Your knees hurt when you exercise or run.
- You feel stiffer and not as flexible as you used to be.
- You notice a little decline in your memory or cognitive abilities.
- You visit your doctor, and he or she tells you your cholesterol or blood pressure is elevated and
- you need to monitor your stress or the foods you're eating—foods that previously you could handle without concern.
- You're diagnosed with cancer.

All of these signs are indications of age-related diseases. For years, the prevailing view has been that, as we age, our body's resources decline through wear and tear, and we become more susceptible to these age-related diseases.

However, there's a growing body of evidence that questions this point of view on aging and actually identifies our mitochondria as the driving force in these age-related issues. That's the aim of this book—to identify the underlying mitochondrial processes that contribute to the aging process. Also pertinent to this discussion are the nutritional, herbal, and lifestyle changes you can make to fundamentally change the conditions that contribute to aging to enhance and prolong your life.

This book will help to educate the reader so you can personalize your approach to the aging process by understanding the foundational concepts of aging. This book will also show how environmental factors within our control can be employed to improve your quality of life.

The implication is that you can improve the quality of your life and extend your lifespan. Improving quality of life means that, as we age, we're still active, both physically and mentally.

First, there are three foundational concepts I want to identify and discuss prior to the discussion of mitochondria and nutritional protocols. These concepts come up over and over in the current research. To get the full value of this book, you can't depend solely on the nutritional protocols; you will need to incorporate aspects of these other concepts into your daily life too.

The mitochondria foundation is

1. Breathing. Your mitochondria require oxygen to complete the chemical process that provides the energy for your body. Chronic shallow breathing (hypoxia) drives mitochondrial dysregulation, which is implicated in age-related diseases. Most notably, chronic shallow breathing occurs when sitting at your desk all day at work, followed by coming home and slumping into your easy chair for a couple of hours of watching TV.

2. Caloric excess and caloric restriction. A definition of caloric excess, in the simplest terms, means eating more calories than your body expends. Those calories are, in large part, composed of carbohydrates and sugars. Sugars and carbs, as we'll discuss in later chapters, drive an energy-production pathway that bypasses the mitochondria and causes an inflammatory response. For a more detailed discussion of the energy value of different calories, see Appendix 2.

3. Exercise is pivotal for mitochondrial biogenesis (growth) (1). As we'll discuss in this book, people who exercise regularly have many more mitochondria than people who have a sedentary lifestyle. Reduced mitochondrial content and functionality may contribute to the onset and/or severity of type 2 diabetes, sarcopenia (age-related loss of skeletal muscle), Alzheimer's disease, and ALS (amyotrophic lateral sclerosis, also called Lou Gehrig's disease) (2). Exercise is a promising treatment for patients with mitochondrial dysfunction.

Deeper breathing, sufficient exercise, and calorie restriction are three controllable environmental factors that have a major impact on the rate of aging and allow us to increase our health span—the number of healthy years we live in a disease-free state.

Mitochondrial oxidative damage is a basic mechanism of aging, and multiple studies demonstrate that this damage is lessened by calorie restriction. The mechanism for sparing mitochondria from damage is based on the observation that calorie restriction mimics cellular hypoxia (low cellular oxygen) and activates a number of hypoxia survival genes. Calorie restriction also increases endogenous antioxidant enzyme levels, which decline with age. These decreased enzyme levels help explain why mitochondria become dysfunctional and damaged during aging and how calorie restriction can reverse this downward trend.

Calorie restriction sends signals to the body that food is becoming scarce, and this activates other survival genes that allow mitochondria to increase their repair rate to fix earlier damage to their DNA and membranes.

These experiments have been duplicated in a wide variety of research animals over the decades,—including chimpanzees, our closest primate relatives—with identical results (3).

Aerobic exercise also mimics hypoxia temporarily by creating an acute need for more oxygen so mitochondria can keep up with energy demands.

Strenuous exercise also temporarily increases free-radical levels in mitochondria, and this has the effect of increasing mitochondrial antioxidant enzyme levels that protect mitochondria long after exercise has stopped (1, 2).

Obesity is a worldwide health epidemic, and practicing calorie restriction is an excellent way to restore humans to a healthy weight and a normal body mass index (BMI). Maintaining normal weight and exercise—as well as drinking alcohol only in moderation, if at all, and not smoking—are four key health factors that we ourselves can control.

Hereditary factors are beyond our control but not as far beyond our control as we once thought, as we'll see in later chapters.

Exercise and maintaining a normal weight, normal blood pressure, and normal blood-serum glucose levels are all directly interrelated. If you perform strenuous aerobic exercise for fifteen to thirty minutes every day or every other day, with a normal daily caloric intake of 2,000 to 2,500 calories, you'll lose excess weight; maintain blood pressure in normal, healthy ranges; burn excess blood glucose; and lower your risk factors of dying from many of the degenerative diseases associated with aging.

All-cause mortality (death from any cause) will also drop, and this will place you in the top 10 percent of the healthiest category of individuals—those least likely to die from cardiac fibrillation and sudden death. Well-exercised, fully oxygenated hearts don't go into fibrillation or suddenly stop beating. In terms of mitochondrial function, excess electrons and protons are properly disposed of during exercise instead of building up in the body to wreak havoc, which will be discussed later in more detail.

Exercise elevates messenger and transfer RNA levels to build and repair muscles, stimulating mitochondrial biogenesis—the process in which new mitochondria are created. Runners and other high-aerobic exercisers have up to double the number of mitochondria in their heart muscles, for example, to increase their energy.

Calorie restriction to maintain normal BMI, coupled with strenuous aerobic exercise, addresses two of the three immediate risk factors we can control.

These three factors can be practiced as routine lifestyle changes by anyone who's motivated to live a healthier and longer life. This book will show you how to combine these three factors with nutritional protocols that will further support and help correct a previously unhealthy lifestyle by putting them to work for you.

Origin of Mitochondria

The origin of mitochondria is essentially the origin of life. Mitochondria were created when one bacterial cell engulfed another, and the resulting independent mitochondria was specialized to become an essential component in the host cell.

I want to take a moment and repeat this statement: The origin of the symbiotic mitochondria relationship is essentially the origin of life. This is profound, and, without this event, we wouldn't be here. This event isn't just a single cellular event; rather, it's an evolutionary milestone that allowed for more complex life forms to develop.

Virtually all life we see around us—plants and animals—are eukaryotes (single-celled or multicellular organisms that contain a distinct, membrane-bound nucleus). Eukaryotes are more complex than prokaryotes (organisms lacking a distinct, membrane-bound nucleus). Eukaryotes contain mitochondria, the energy-producing organelles, inside their cell walls. Eukaryotic cells eventually evolved into multicellular organisms, and this was the enormous step forward that was necessary for multicellular evolution (4).

During evolution, mitochondria initially existed as separate bacterial cells (related to today's disease-causing blue bacteria) that were absorbed by prokaryotes in a symbiotic, mutually supportive role in which a division of labor occurred. Mitochondria provided the energy for the cell, and the cell provided a safe harbor of nutrition inside its cell wall. This process of mutual support by the fusion of cell types is called endosymbiosis (6).

If mitochondria hadn't been incorporated into single-celled organisms as an energy source, life wouldn't have advanced beyond the stage of bacteria. Bacteria are prokaryotes—organisms that have no nucleus, no well-defined organelles inside their cell walls, and possess no mitochondria. The incorporation of an internal power source (mitochondria) into the cell (eukaryotic) allowed for the evolutionary advancement of complex, multicellular organisms.

Bacteria passively obtain energy when nutrients travel through their cell walls. All food for these single-celled prokaryotes must be absorbed from the outer membrane, and the organism can only grow so large before it collapses due to the forces of gravity and other cell-membrane integrity issues (5).

It's recognized that evolution typically occurs in ecological boundary zones, such as where freshwater merges with seawater or deep in the ocean where hot sulfur vents interface with cold seawater. In these settings, we have planet-wide biological experiments occurring on a massive scale, pushing the single-celled organisms up the evolutionary ladder. This could include a sulfur breather from a deep ocean vent engulfed an oxygen breather or an oceanic cellular organism engulfed a freshwater cell.

Basic Facts about Mitochondria

Mitochondria are the primary source of energy in the cell, and mitochondria keep the cell alive. Mitochondria are the only organelles in the cell that have a dual-walled membrane—they generate energy for cells by reducing oxygen to water in a four-step reduction to produce adenosine triphosphate (ATP). ATP is present in all cells, where it's used to store and transport energy needs for biochemical reactions. In the course of this process, oxidants are produced by normal mitochondrial metabolism (7). ATP and its critical function are described in detail in Chapter 2.

Structure of Mitochondria

Mitochondria are ideally shaped to maximize their role in energy production. They have an ordinary outer membrane like other organelles inside the cell. They have an inner membrane that has numerous folds in it, called cristae, which provide much greater surface area to enhance the productivity of cellular respiration.

There's an intermembrane space between the inner and outer membranes that has the same composition as the interior of the cell except that its protein content differs.

The matrix of the mitochondria is the inner fluid inside the two membranes and comprises a complex mixture of enzymes and other proteins. These enzymes are important for the synthesis of ATP and also play a role in the maintenance of mitochondrial ribosomes, transfer RNAs, and mitochondrial DNA (mtDNA) (8).

Composition in Human Body

There are hundreds to thousands of mitochondria present in each cell of the body. Each cell contains hundreds of mitochondrial nuclei, with five to ten copies of the genetic code in each nucleus. This is a much larger volume of mitochondria DNA (mtDNA) than nuclear DNA (two copies per one nucleus per cell).

MtDNA is necessary because cells are constantly responding to energy demands on a cell-by-cell basis, requiring the presence of controlling DNA near the source of energy production. Mitochondria produce more than 90 percent of the body's energy.

Cells require tremendous amounts of energy to function and support cellular structure, regeneration, and growth. Cellular respiration occurs predominantly in the mitochondria. During this process, cells ingest nutrients (glucose and fatty acids) that are broken down to produce valuable energy compounds (ATP is the primary energy compound in the body) that are relayed to cells. ATP has high-energy bonds that hold the phosphate groups together; once broken by ATPase-degrading enzymes, these bonds release large amounts of energy. The pool of ATP must be regenerated at adequate levels to supply the body.

Mitochondrial Number in Cell Types

There are approximately one hundred to eight thousand mitochondria in each cell, depending on how much energy the cell needs to produce to carry out its function. Each cell type develops and maintains a specific mitochondrial capacity to satisfy its metabolic and energetic demands. The division of the mitochondria is a result of the energy demand, so the cells with the highest energy must have the greatest number of mitochondria.

Muscle cells, including cardiac tissues, have a higher mitochondrial content—because of their higher energy needs—than other, less metabolically active cells. Other cells and tissues with high metabolic activity include liver and fat cells.

Cell types with a lower mitochondrial content include lymphocytes, macrophages, and other immune responder cells. Red blood cells entirely lack mitochondria because they're simply empty carrier cells (9).

The number of mitochondria can be increased in cells in a process called mitochondrial biogenesis, which is triggered by a process called oxidative stress. Oxidative stress is created by an imbalance among free-radical production, endogenous antioxidant enzyme levels, and the amount of dietary antioxidants like vitamins C and E (10).

ATP Energy Output

The primary function of mitochondria is ATP production. The ATP molecule is somewhat unstable in that it's easily converted into ADP (adenosine diphosphate) by removing one of the phosphate-oxygen groups. The ATP released from energy production cycles in the mitochondria can be easily used by the other components inside the cell. Aerobic respiration produces ATP by a process called oxidative phosphorylation, or OXIPHOS, where glucose is oxidized in three stages:

1. The first stage is called glycolysis, where oxygen oxidizes glucose in the cell outside the mitochondria to produce two molecules of glucose and pyruvate.
2. Pyruvate then enters the citric acid—or Kreb's—cycle to produce some ATP and CO_2 inside the mitochondria.
3. Finally, the oxidation of pyruvate produces acetyl coenzyme A, and the energy of acetyl coenzyme A is harvested to produce most of the ATP in a four-stage process of electron transfers to oxygen that reduce it to water. This step is called mitochondrial respiratory chain Complex I–IV.

The mitochondrial respiratory chain transfer of electrons isn't perfect—there's a 1 percent leakage of electrons from mitochondrial respiratory stage I, and a smaller leakage of electrons from stage III (11). This electron leakage reacts with the available oxygen inside all cells to produce superoxide radicals. In later chapters, we'll show why this electron leakage is an important cause of aging and deterioration of the cells.

Glucose is the most commonly used carbohydrate for fuel and is always used as the first choice to make energy. Fatty acids are used as a second choice for energy when glucose is in short supply. Rarely, amino acids are oxidized for energy as a last resort.

One glucose molecule produces two ATP molecules (some of the ATP is used up in the process). This is a quick-and-dirty method for producing ATP—"quick" meaning the intake of carbohydrates and sugars is high, and "dirty" meaning the process produces free radicals—which, as we'll discuss, have a detrimental effect on the body.

Fatty acids, on the other hand, are stored as triglycerides in the fat cells of animals and humans. Triglycerides are oxidized differently, in a process called beta-oxidation—also known as B-oxidation—by mitochondria in the citric acid cycle to yield CO_2 and water along with thirty-six ATP molecules (12).

Free Radicals and Oxidants

The energy production inside mitochondria comes at a cost—free radicals are released, which are implicated in aging and the development of cancer and other degenerative diseases in multicellular organisms. This can be directly traced back to mitochondrial activity.

A free radical is an atom or group of atoms that has at least one unpaired electron and is therefore unstable and highly reactive. When a chemical reaction is over, some molecules end up with unpaired electrons. Electrons travel in pairs—an alpha electron coupled with a beta electron—around the nucleus. These are analogous to a magnetic field, which has a north pole and a south pole. Electrons are powerful electromotive forces that repel each other when they're both of the same spin, but their repellant forces cancel each other out when they're paired properly. This pairing provides stability to most atoms and molecules.

We depend on the generation of free radicals for signaling and enzyme activity inside the cell, and there's no question that free radicals are necessary for life itself. The great majority of chemical reactions proceed with the production of free-radical species in intermediate stages of the reaction (12). Thus, the free radicals produced from chemical reactions necessary to keep us alive are a small price to pay for performing the body's essential functions, such as breathing or generating energy.

The problem with free radicals produced for normal physiological processes is that they drift far beyond the site where they're generated and damage DNA and other body proteins. Antioxidants, on the other hand, donate electrons or hydrogens to pair up the unpaired electrons in the orbits of free radicals. Antioxidants are usually electron or hydrogen donors, and free radicals usually act as electron or hydrogen acceptors (12).

References:

1. Wang, L., Mascher, H., Psilander, N., Blomstrand, E., and Sahlin, K. "Resistance Exercise Enhances the Molecular Signaling of Mitochondrial Biogenesis Induced by Endurance." *Journal of Applied Physiology* 111, no. 5 (November 2011): 1335–44.

2. Russell, A.P., Foletta, V.C., Snow, R.J., and Wadley, G.D. "Skeletal Muscle Mitochondria: A Major Player in Exercise, Health, and Disease." *Biochimica et Biophysica Acta* 1840, no. 4 (April 2014): 1276–84.

3. Anton, S. and Leeuwenburgh, C. "Fasting or Caloric Restriction for Healthy Aging." *Experimental Gerontology* 48, no. 10 (October 2013): 1003–5.

4. Baum, D. and Baum, B. "An Inside-Out Origin for the Eukaryotic Cell." *BMC Biology* 12:76 (2014).

5. Nace, G.W. "Gravity and Positional Homeostasis of the Cell." *Advances in Space Research* 3, no. 9 (1983): 159–68.

6. Martin, W.F., Garg, S., and Zimorski, V. "Endosymbiotic Theories for Eukaryote Origin." *Philosophical Transactions of the Royal Society* (August 31, 2015).

7. Frey, T. and Mannella, C. "The Internal Structure of Mitochondria." *Trends in Biochemical Sciences* 25 (July 2000).

8. King, M.W., "Mitochondrial Functions and Biological Oxidations." 1996–2016. themedicalbiochemistrypage.org, LLC.

9. Robin, E. and Wong, R.J. "Mitochondrial DNA Molecules and Virtual Number of Mitochondria per Cell

in Mammalian Cells." *Journal of Cellular Physiology*, 136, no. 3. (September 1988): 507–13.

10. Jomayvaz, F. and Shulman, G. "Regulation of Mitochondrial Biogenesis." *Essays in Biochemistry* 47 (June 14, 2010): 69–84.
11. Czura, A.W. "Energy Production in a Cell: Chapter 25 Cellular Respiration Handout." Lecture materials from Suffolk County Community College BIO130.
12. Pryor, W. "Chapter One," *Free Radicals in Biology*, 1st ed. (Amsterdam, Netherlands: Elsevier, 1976).

Chapter 2

Molecular Machines in the Cell

"All self-organizing systems are wholes made up of parts, which are themselves wholes at a lower level, such as atoms in molecules and molecules in crystals." (1)

Before we begin the journey of the mitochondria, it's important to understand the cellular landscape we're traversing. All life started out billions of years ago as the dancing of molecules. This dance today is most easily observed at the cellular level as groups of molecules that have come together to form molecular machines. As you'll learn, the mitochondria are molecular machines—little powerhouses. Molecular machines transport nutrients, expel waste, make new machines, and read and translate DNA.

Probably when you were in high school or college, you had to take biology, and your view of the cell was based on an image similar to this:

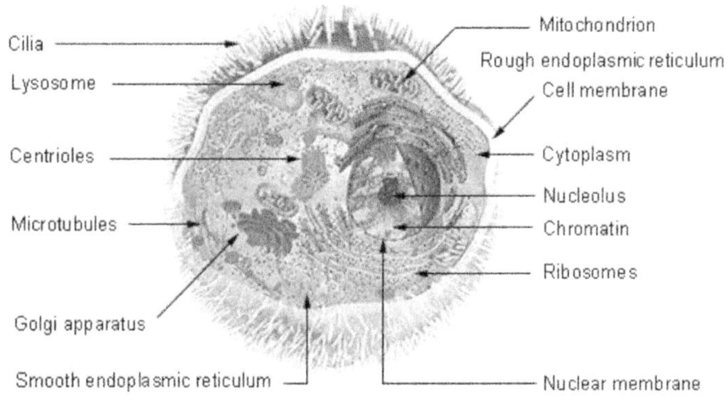

Figure 1. Structure of the Cell
Credit:
https://commons.wikimedia.org/wiki/File:Illu_cell_structure.jpg

The image above appears somewhat orderly. You may think the cellular environment is calm and quiet and that most of the chemical processes occurring in the cell are of the passive, osmotic type, but that view doesn't even come close to describing the remarkable processes occurring and the reality of the environment at the cellular level.

With the development of the scanning electron microscope, we're now able to view biological structures at sub-nanometer resolution—at the level of the molecules—and we can see the intricate, complex structure of molecular machines.

For example, here's an image of a section of a mitochondrial dome taken with a scanning electron microscope, enhanced with computer imaging software and colorized for easier viewing:

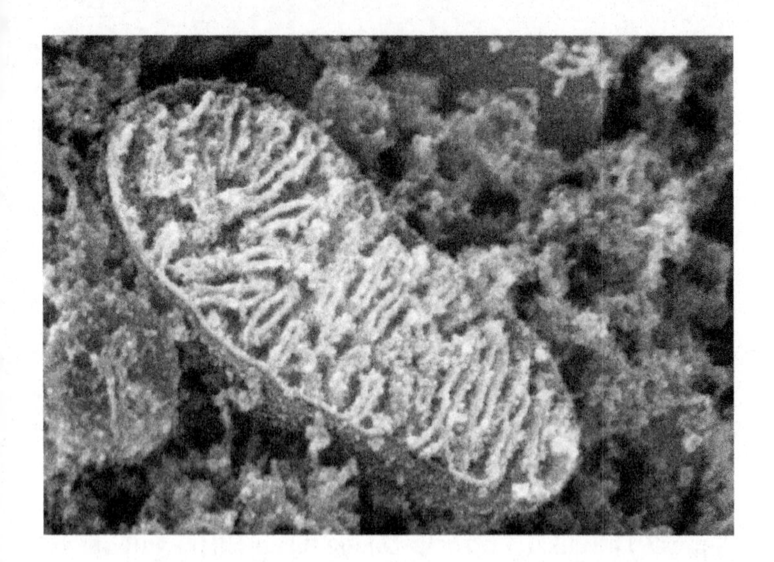

Figure 2. Electron Microscope View of the Mitochondrial Dome
Credit: https://www.keele.ac.uk/lifesci/people/davefurness/

As the image above shows, cells are incredibly crowded places, stuffed full of large proteins, DNA and RNA molecules, sugars, lipids, ions, and innumerable water molecules. It's been estimated that the average space between proteins in living cells is less than ten nanometers. Proteins are between ten and one hundred nanometers in size. In more relatable terms, this is equivalent to a crowded parking lot with just a foot or less space between each car. When things are this tight, it becomes tricky for components to maneuver past one other. In addition to this crowding, every space between proteins is filled with water, ions, sugars, and other assorted small molecules (2).

The cell as a whole can be viewed as a functioning city, populated by various buildings that represent important structures within a cell. There's a library (the nucleus), which contains the genetic material; power plants (mitochondria); highways (microtubules and actin filaments); trucks (kinesin and dynein); garbage disposals (lysosomes); city walls (membranes);

post office (Golgi apparatus); and many other structures fulfilling vital functions. Molecular machines perform all of these functions. Some machines twist DNA; some route cargo along molecular highways or through the cell membrane. There are numerous such machines in living cells, doing many different things, moving in different ways, working together in ways we still don't fully understand (3).

Some molecular machines, such as kinesin and myosin, are involved in moving materials to different parts of the cell or rebuilding the microtubes that support the cell structure. Remember our discussion in Chapter 1 about the prokaryotic cells, which can only grow so big before the cell membrane collapses? The eukaryotic cells handle this problem with internal structure to support the outer cellular membrane.

Kinesins are motorized transport machines that move cellular materials—including mitochondria—to where they're needed in the cell so they can perform their functions. Kinesins have two feet, or "globular heads," that literally walk, one foot over another. Known as the workhorses of the cell, kinesins can carry cargo many times their own size. Here's a great video showing a kinesin "walking" and pulling cargo:

https: www.youtube.com/watch?v=gbycQf1TbM0
Courtesy: www.evolutionnews.org

A cell generates waste. Carbon dioxide and urea, the byproducts of energy production, are expelled and disposed of elsewhere. Many components of the cell eventually wear out and need to be broken down and have their parts recycled. Molecular machines are involved in this recycling and moving materials to be recycled. This activity takes place inside the cell in specialized compartments called lysosomes, which are fluid-filled "bags" of membrane that contain a cocktail of molecules for breaking down complex molecules and providing the cell with the simpler nutrients it requires. A mitochondriom that has passed its sell-by date, for example, is engulfed, disassembled, and reused by the cell. The beauty of the cell is that most of its waste is recycled. Our cells offer an elegant example of how our planet could be better managed.

The environment of this tiny cellular city is equivalent to being in a hurricane. Let me explain. Water is unlike any other liquid in the universe. For example, compared with liquids of similar molecular construction, water has to be heated to a much higher temperature for it to melt or boil. It also has the curious property that its solid form (ice) is lighter than its liquid form; this is why ice cubes in your drink float rather than sink.

At the nanoscale level, water is in continuous motion. Hydrogen bonds arise in each water molecule—which is made up of one oxygen atom and two hydrogen atoms—because the oxygen "steals" the hydrogen's electrons. The leftover hydrogen ions are positively charged while the oxygen atom, now having two excess electrons, is negatively charged. When two neighboring water molecules come close to each other, the positively charged hydrogen on one molecule is attracted to the negatively charged oxygen on the neighboring water molecule. In liquid water, these hydrogen bonds form and break continuously at a high rate (4).

Consider this image: You're sitting on the bank of a lake. It's a quiet and sunny day, and the lake is almost reflective. If you could see the lake surface at a higher magnification, you'd observe a mist over it. If the magnification of this image is increased to the nanoscale level, the lake surface disappears; all that remains is a cloud moving violently, as if boiling. Atoms and molecules are restless. In air, molecules of nitrogen, oxygen, carbon dioxide, and water vapor randomly swirl around, colliding at high speeds. The calmness we see around us is merely an illusion.

At the cellular level, the molecular storm (the boiling cloud) reigns supreme. The tiny scale of atoms and molecules is dominated by continuous motion. Scientists call this continuous motion of atoms and molecules thermal motion. Thermal motion does not mean the atoms are floating gently, however. At room temperature, air molecules reach speeds in excess of the fastest jet airplane! If we were reduced to the size of a molecule, we'd be bombarded by a molecular storm—a storm so fierce, it would make a hurricane look like a breeze. If you think the molecular storm is amazing, consider that, in a millisecond, an average molecule undergoes ten billion collisions with water molecules.

Lipids are another group of molecules that can organize into molecular machines and form more complicated cooperative structures in water, such as a cell membrane. Cellular membranes are double-walled spheres called vesicles, which are composed of lipids (fatty acids). Inside the double wall, the hydrophobic (water-repellant) sides are safely tucked away from the surrounding water; the two surfaces, one on the outside and one on the inside of the sphere, face water molecules. A vesicle separates one volume of water from another. If we place chemicals inside the vesicle, we create an isolated, nanosized reaction chamber. Imagine a cell floating in an extracellular environment. Many complex processes occur within the cell that require homeostasis (internal equilibrium). The cellular membrane composed of lipids protects this interior environment from the extracellular environment.

The Golgi body, discovered in 1897 by Camillo Golgi, is another molecular machine located within the cell, as is the endoplasmic reticulum. These lie in the cytoplasm outside the nucleus, like the mitochondria. Their function is to sort, package, label, and modify proteins and lipids, and then transport the finished products to different parts of the cell and through the cell wall to other places. They act much like the post office—but, fortunately, are more efficient.

Enzymes are also remarkable molecular machines. Consider an enzyme called phosphoglucomutase. This enzyme converts an indigestible kind of sugar into one more usable by the body. By speeding up the conversion process by a factor of one trillion, a single phosphoglucomutase molecule can convert a hundred sugar molecules per second. Without this enzyme, it would take three hundred years to convert just one sugar molecule (5)!

Molecular machines need a supply of energy, which comes from the food we eat. As part of metabolism, enzymes residing in the stomach, intestines, and cells break food down. The substrate (substances on which enzymes act) nutrients—carbohydrates, fats, and proteins—are used by the mitochondria in a complex process to make ATP, the energy storage molecule that brings the molecular machines to life.

Three phosphates (triphosphate) bind to adenosine to form adenosine triphosphate, or ATP. With all three phosphates attached, ATP is a bundle of concentrated energy. Snapping off one or two of the phosphate groups releases a great deal of energy—only the molecule's activation barrier keeps the phosphates from detaching right away. But once ATP binds to a molecular machine, the phosphate groups snap off readily, ATP turns into ADP (adenosine diphosphate), and the machine is provided with a large amount of energy. The energy released by the loss of one phosphate is equivalent to heating the enzyme up to seven thousand degrees Fahrenheit (6).

Through the addition of water, molecular motors turn ATP into ADP. They then release ADP, and the spent molecular fuel floats away. The cellular environment is very keen on recycling; the

ADP is recharged by reattaching a phosphate and turning it back into ATP. This is accomplished by one of the more amazing molecular machines, ATP synthase, which is a rotary machine located within the mitochondria. Here's a video on how ATP synthase works:

https://www.youtube.com/watch?v=XI8m6o0gXDY
Credit: DiscoveryScienceNews

Cells have a rather impermeable membrane, but this membrane is pockmarked with a vast array of specialized pores. Cellular homeostasis is extremely important for the cell's survival, and the cell needs ways to control what's coming in and going out. Many of the pores are passive: They let certain molecules through if the molecules happen to diffuse that way. But some are machines—active pumps that move certain ions or molecules into or out of the cell—and require the energy generated by ATP. This is discussed in greater detail with a focus on cancer, including how these active pumps can be closed to kill the cancer cell, in Chapter 10.

Usually, ions will move across a membrane until they reach equilibrium. Passive pores will let ions diffuse, but the ions will only diffuse until equilibrium is reached. To create a nonequilibrium concentration difference between the inside and the outside of the cells, active pumping is required. One such pump, the sodium-potassium pump, moves sodium out of the cell and potassium into the cell, maintaining the typical high-potassium, low-sodium environment found in most cells (7). These pumps are extremely important for the cell, and our cells typically expend one-third of their energy to run these sodium-potassium pumps to maintain a proper balance. For a demonstration of this process, you can view these videos:

https://www.youtube.com/watch?v=bFonzS22xN8
Credit: http://www.mheducation.com/permissions.html

Molecular machines are highly complex and, in many cases, we're just beginning to understand their inner workings. What follows is a partial list that briefly describes five molecular machines identified in the scientific literature.

1. Eukaryotic cilium: The cilium is a hair- or whip-like structure built on a system of microtubules, typically with nine outer microtubule pairs and two inner microtubules. These machines perform many functions in eukaryotes, such as allowing sperm to swim.

https://www.youtube.com/watch?v=vQ3CdSiVzUk
Credit: Virtual Lung Project,
http://www.youtube.com/user/TheCISMM

2. Ribosome: The ribosome is a ribonucleic acid (RNA) machine involving more than 300 proteins and RNAs. It forms an organelle in which messenger RNA is translated into protein, thereby playing a crucial role in protein synthesis in the cell.

https://www.youtube.com/watch?v=TfYf_rPWUdY
Credit: www.dnale.org

3. F0F1 ATP synthase: This protein-based molecular machine, which is part of the larger mitochondrial molecular machine, is actually composed of two distinct rotary motors that are joined by a stator (nonmoving part of the rotor). As the F0 motor is powered by protons, it turns the F1 motor. This kinetic energy is used like a generator to synthesize ATP.

https://www.youtube.com/watch?v=3y1dO4nNaKY&t=44s
Credit: http://vcell.ndsu.edu/animations

4. Kinesin motor: Much like myosin, kinesin is a protein machine that looks like a stick man. It binds to and carries cargo along a microtubule in the cell. Kinesins, like vesicles, are powerful enough to drag large cellular organelles drag within he cell.

https://www.youtube.com/watch?v=gbycQf1TbM0
Credit: DiscoveryScienceNews, www.discovery.org

5. Spliceosome: DNA sequences held in the cellular nucleus contain the coding of life; however, this DNA is interrupted by stretches of noncoding sequences. The spliceosome removes the stretches of noncoding sequences. This may not sound like much, but consider the importance of how the spliceosome ensures the ongoing integrity of your DNA.

https://www.youtube.com/watch?v=aVgwr0QpYNE
Courtesy: http://www.dnalc.org/

When we think of our bodies, usually we only consider our organs and, when we visit our doctor, we typically only get information about the functionality of these organs. However, if our scale of vision were reduced to the molecular level, we would find a degree of detail and complexity difficult to imagine, which is why I've tried to include numerous visual aids.

The next chapter details the inner workings of that remarkable molecular machine, the mitochondria, which creates the energy that drives the organs and allows *you* to exist.

References:

1. Sheldrake, R. *Science Set Free: 10 Paths to New Discovery* (New York: Crown Publishing Group, 2012), p. 50, Kindle.

2. Hoffmann, P.M. *Life's Ratchet: How Molecular Machines Extract Order from Chaos* (New York: Basic Books, 2012), Chap 4, Kindle.

3. Hoffmann, P.M. *Life's Ratchet: How Molecular Machines Extract Order from Chaos* (New York: Basic Books, 2012), Chap 7, Kindle.

4. Hoffmann, P.M. *Life's Ratchet: How Molecular Machines Extract Order from Chaos* (New York: Basic Books, 2012), Chap 4, Kindle.

5. Hoffmann, P.M. *Life's Ratchet: How Molecular Machines Extract Order from Chaos* (New York: Basic Books, 2012), Chap 6, Kindle.

6. Hoffmann, P.M. *Life's Ratchet: How Molecular Machines Extract Order from Chaos* (New York: Basic Books, 2012), Chap 6, Kindle.

7. Hoffmann, P.M. *Life's Ratchet: How Molecular Machines Extract Order from Chaos* (New York: Basic Books, 2012), Chap 7, Kindle.

Chapter 3

Theory of Aging

"'The Tithonus Error'— the presumption that, when we talk about combating aging, we're only talking about stretching out the grim years of debilitation and disease with which most people's lives currently end." (1)

Aging is a process that results in a decline of optimal body function through a combination of mutant mitochondrial and cellular debris accumulation. In this chapter, we'll demonstrate how a few mutant mitochondria in our cells cause our entire body to age. These mutated mitochondria take over a tiny 1 percent of cells by clonal expansion. We'll show how these rogue mitochondria evade detection by mechanisms designed to identify and eliminate them in the first place (2).

Aging also results in an accumulation of cellular debris caused by oxidized proteins and the cross-linking of proteins in the body by glucose. Excessive consumption of carbohydrates exacerbates this process of glucose cross-linkage, so watching your carbohydrate consumption in terms of quantity and type is wise. For example, carbohydrates with a high glycemic index, such as white rice, pastas, white bread, etc., should be avoided.

The answer to aging involves rejuvenation using seven repair strategies aimed at the seven forms of cellular damage that manifest as aging. Cleaning the junk from cells (inside and out), nullifying havoc from mutations in mitochondria and the nucleus, eliminating protein cross-linking, and replacing lost cells—as well as getting rid of bad cells that are causing problems—is the magic combination for aging gracefully.

Simply stated, old age may simply be the consequence of lax biological housecleaning, which is something we can influence if armed with the right information (2).

Chronic degenerative diseases are strongly associated with advancing age. These include cardiovascular diseases, diabetes, metabolic syndrome, neurodegeneration, and cancer. Although unique in their clinical manifestations, degenerative diseases have one aspect in common—they're driven by oxidative stress from mitochondrial dysfunction, which triggers chronic inflammation throughout the body (2, 3).

Oxidants and free radicals, as we've seen, are generated by normal metabolic processes and are essential to life. However, overproduction of oxidants sometimes overwhelms the body's own antioxidant defenses, creating a condition known as "oxidative stress" (4).

The two types of antioxidant defenses consist of antioxidant enzymes produced internally by the cells and small-molecule antioxidants such as vitamins C and E, which are obtained from the diet (4).

The vast majority of mitochondria remain healthy in old age, but about 1 percent of older people's cells are completely taken over by rogue mitochondria that have defective mitochondrial DNA. Studies have revealed that these rogue mitochondria all have exactly the same defects or mutations in their DNA! How could that be?

Far from being a coincidence, cells that have identical defective mutations in their mitochondrial DNA can be a driving force in the aging process, contributing to inflammation throughout the body. This inflammation can cause, for example, LDL (low-density lipoprotein, so-called "bad" cholesterol) oxidation, the earliest step in hardening of the arteries, the medical term for which is atherosclerosis (2).

Atherosclerosis is one example of how mutant mitochondria can cause disease in distant parts of the body—namely, the entire circulatory system. Atherosclerosis is the main cause of cardiovascular disease, which is the leading cause of death in humans (5). So, clearly, this is an important health consideration.

Energy is created by mitochondria using virtually identical principles to those used by hydroelectric dams. Electrons run a series of pumps called the electron transport chain, which fill a reservoir of protons held back by a dam—the inner mitochondrial membrane (2). The turbine that generates electricity is driven by the downward flow of protons, like water flowing downhill.

The inner membrane contains a turbine, called Complex V, which is driven by this flow of protons. Converting ADP into ATP (which occurs by recharging ADP by reattaching a phosphate), the energy storage currency of the cell, harnesses the protons rushing through Complex V. Unlike a hydroelectric dam, this energy flow isn't a structural process driven by gravity and force but a chemical process (2).

Sure enough, the mitochondrial inner membrane and the mitochondrial DNA are the first targets of a superoxide radical attack. Superoxide, as a highly unstable molecule with an unpaired electron, desperately tries to find an electron to pair with its unpaired electron, so it steals an electron from the mitochondrial DNA, resulting in a mutation. The unpaired electron makes superoxide, which is very toxic biologically because it steals an electron from another compound upon contact. When superoxide collides with the membranes of the mitochondria, it steals an electron from the easily oxidized fatty acids that make up the membrane, causing a free-radical cascade known as propagation (5). Ultimately, superoxides may contribute to the pathogenesis of many diseases.

Propagation causes eight to ten molecules along the mitochondrial membrane to become damaged before this runaway freight train is brought under control (6). This results in damaged, poorly functioning mitochondrial membranes that have trouble passing carnitines across them to generate ATP.

At some point, this damage builds up to the point where cytochrome c is released by the inner mitochondrial membrane. The leakage of cytochrome c triggers the beginning of a process known as apoptosis, a form of cellular suicide.

Apoptosis is believed to be a built-in safety method that allows badly damaged mitochondria and other cells to self-destruct before they begin to become a threat to the living organism as a whole. These threats come in the form of cancers or systemic damage caused by rogue mitochondria (2, 6).

This self-destruction process only weeds out mitochondria with damaged membranes. Mitochondria with intact membranes—but damaged DNA—wouldn't show any outward signs of their internal injuries so would be passed over by this self-destructive apoptotic process (2).

Mitochondria also evade detection from lysosomes, which are another type of organelle found in cells that digest cellular debris using enzymes that break down large proteins produced during aging. Lysosomes detect badly damaged mitochondrial cell membranes by the chemicals they give off and attack and dissolve these mitochondria (7). But mitochondria that have badly damaged DNA and intact outer membranes don't give off chemo-attractants and are spared from destruction by the lysosomes (2).

How Cells Taken Over by Mutated Mitochondria Escape Detection

Cells that have been taken over by mutant mitochondria survive and gradually accumulate with age. These cells evade cellular suicide, or apoptosis, by switching their ATP energy production to a process called anaerobic glycolysis, which differs greatly from the normal process of ATP production—aerobic glycolysis (2).

Normal aerobic glycolysis is the metabolism of glucose from food, and it takes place in the main body of the cell. Glycolysis generates a small amount of ATP, a metabolite called pyruvate, and some electrons, which drive oxidative phosphorylation in the mitochondria (8).

Electrons are loaded onto a carrier molecule called NAD+ to gain entry into the mitochondria. The addition of an electron to NAD+ converts the molecule to NADH (nicotinamide adenine dinucleotide). The pyruvate formed is also carried into the mitochondria, where it's changed into another intermediate called acetyl coenzyme A (acetyl-CoA). This process releases more electrons harvested by charging NAD+ into NADH (9).

Acetyl-CoA is used as the raw material for a series of reactions called the tricarboxylic acid cycle (TCA cycle), or Krebs cycle, which liberates many more electrons and creates much more NADH than the previous steps.

Finally, all the NADH charged up by these steps is fed into the electron transport chain in the mitochondria, which uses this electron "payload" to generate the proton reservoir that drives the generation of nearly all the cell's energy (2, 9).

On the other hand, cells that have mutant mitochondria have an unusually active TCA cycle that's in overdrive, which allows these cells to double their production of ATP from glucose because of an electron-exporting feature called the plasma membrane redox system, or PMRS (2). This system allows cells to export electrons from NADH inside the cell and transport them outside the cell. This export allows even normal cells to have better control over the reducing and oxidizing factors inside them and keep tighter control over the availability of NAD+ and NADH.

In cells with mutant mitochondria, PMRS allows the cells to abandon aerobic glycolysis and still get the ATP they need. The drastic increase of PMRS, with its greatly increased TCA activity, would make the cell surfaces of mutant mitochondrial cells absolutely overflow with electrons from this export process, forming a hotspot of negative electrical charge due to the surplus of electrons (2, 10). Oxygen is the only molecule in great abundance in bodily fluids that can sponge up these excess electrons that mutant mitochondrial cells are capable of generating.

When oxygen takes up an extra electron, it's converted into a superoxide radical. A superoxide radical is highly reactive, damaging the first nearby cell membrane or protein it comes into contact with. This would be an especially dangerous situation if superoxide were to come into contact with particles that travel all over the body, such as LDL cholesterol particles (2, 11). LDL cholesterol docks at endothelial docking spaces to deliver cholesterol all over the circulatory system. The endothelium is the outer lining of the arteries and other circulatory cells. Cholesterol is used for cell membrane construction and other structural purposes.

Because LDL cholesterol has fatty-acid components that are easily oxidized, superoxide could preferentially oxidize a fatty acid, leading to the propagation of free radicals (6). Put in simple terms, oxidized LDL cholesterol can be compared to rancid oil. This rancid oil forms harmful free radicals in the body, which are known to cause cellular damage and have been associated with the development of diabetes, Alzheimer's disease, and other degenerative conditions. LDL cholesterol can also cause digestive distress and deplete the body of vitamins B and E.

Oxidized LDL cholesterol is no longer recognized by the endothelial-cell recognition system. Instead, it's targeted by the macrophages as a foreign invading protein, so the macrophages engulf it in a process called phagocytosis. The resulting fused product binds to the endothelial cell to produce a foam cell, which is the first step in atherosclerosis (11).

This is one example of how mutant mitochondrial cells can spread oxidative stress throughout the body to wreak serious health consequences (2). This process could be a central driver of systemic biological aging.

In the next chapter, we'll investigate the function of mitochondria in greater detail and how inefficiencies in electron transport within mitochondria result in the creation of free radicals, which are implicated in the development of a number of degenerative diseases associated with aging.

References:

1. de Grey, A. and Rae, M. Ending Aging: *The Rejuvenation Breakthroughs That Could Reverse Human Aging in Our Lifetime* (New York: St. Martin's Press, 2007), p. 8, Kindle.
2. de Grey, A. and Rae, M. Ending Aging: *The Rejuvenation Breakthroughs That Could Reverse Human Aging in Our Lifetime* (New York: St. Martin's Press, 2007), p. 59, Kindle.
3. Shigenaga, M.K., Hagen, T.M., and Ames, B.N. "Oxidative Damage and Mitochondrial Decay in Aging." *Proceedings of the National Academy of Sciences (PNAS) of the United States of America* 91, no. 23 (November 1994): 10771–8.
4. Sies, H. *Oxidative Stress* (London: Academic Press, 1985).
5. Lusis, A.J. "Atherosclerosis." *Nature* 407, no. 6801 (September 14, 2000): 233–41.
6. Elmore, S. "Apoptosis: A Review of Programmed Cell Death." *Toxicologic Pathology* 35, no. 4 (2007): 495–516.
7. Luzio, J., Pryor, P., and Bright, N. "Lysosomes: Fusion and Function." *Nature Reviews Molecular Cell Biology* 8, no. 8 (August 2007): 622–32.
8. Lunt, S.Y. and Vander Heiden, M.G. "Aerobic Glycolysis: Meeting the Metabolic Requirements of Cell Proliferation." *Annual Review of Cell and Developmental Biology* 27 (2011): 441–64.
9. Berg, J.M., Tymoczko J.L., and Stryer, L. "The Citric Acid Cycle Oxidizes Two-Carbon Units," Section 17.1, in *Biochemistry*, 5th ed. (New York: WH Freeman, 2002).
10. Kowald, A. and Kirkwood, T. "Evolution of the Mitochondrial Fusion–Fission Cycle and Its Role in

Aging." *PNAS* 108, no. 25 (2011): 10237–42.

11. Levitan, I., Volkov, S., and Subbaiah, P. "Oxidized LDL: Diversity, Patterns of Recognition, and Pathophysiology." *Antioxidants and Redox Signaling* 13, no. 1 (July 1, 2010): 39–75.

Chapter 4

Mitochondria and How They Make Energy

Energy Generation, Cell Health, and Metabolism

Efficient energy generation is the key to understanding the difference between healthy cells, senescent (aging) cells, and diseased cells. Energy is the cornerstone of all life.

The process of energy generation and utilization in living things is called metabolism, which is the sum of all chemical reactions that take place in each cell and provide energy for essential processes, including the synthesis of new organic compounds. Metabolism and its efficiency in producing cellular energy is the limiting factor in determining the quality of health of a cell, a tissue, or a whole person. Optimal metabolism results in peak performance of an individual when his or her energy output is at its maximal level.

Conversely, a person who has suboptimal cellular energy production will experience less vitality, declining health, and a greater susceptibility to infections and diseases.

As we saw in the last chapter, mitochondria are specialized structures unique to the cells of humans, animals, plants, and other eukaryotes that create energy for cellular function. They serve as energy generators, supplying power for various cellular functions and for the organism as a whole.

There's a direct relationship between mitochondrial health, energy production, and diseases in the whole spectrum of eukaryotes, from the least complex multicellular organisms to humans.

In virtually every disease, mitochondrial dysfunction is at the center of events and is usually the origin and driving force behind the pathology.

Loss of energy production can be the result of inherited mitochondrial defects, somatic damage during aging, or a combination of these factors. We first need a detailed understanding of normal mitochondrial function to understand how mitochondria become dysfunctional.

The Structure of Mitochondria

Mitochondria are the only organelles in the cell that have a dual membrane. The outer membrane is smooth, but the inner membrane, the cristae, is folded to increase its surface area to accommodate increased energy production. Key enzymes and electron-transfer molecules are found in the cristae itself. The space between the two membranes is called the matrix, which contains enzymes needed to facilitate the transfer of energy compounds from the Krebs cycle to the electron transport chain. These two processes will be described later in detail.

Composition and Numbers

Ten percent of the human body is made up of mitochondria—totaling about ten million billion. Between one thousand to two thousand mitochondria are present in each cell. The mitochondria contain their own DNA—thirteen DNA genes to be exact. During evolution, the vast majority of mitochondrial genes were transferred to the safe harbor of the cell's nucleus to protect them from the massive free-radical production that occurs in the mitochondria itself. It was necessary to retain some mitochondrial DNA, however, because cells are constantly responding to energy demands on a cell-by-cell basis. This requires the presence of controlling DNA near where energy is produced. The mitochondria produce more than 90 percent of the energy of the organism in which they reside. In this context, cells require tremendous amounts of energy to function and support cellular structure, regeneration, and growth.

Cells ingest nutrients, mostly glucose and fatty acids, which are broken down to produce valuable energy compounds, in particular, ATP, the universal energy compound of the cell. ATP has high-energy bonds that hold the phosphate groups together. When these bonds are broken by ATPase enzymes, large amounts of energy are released. The pool of ATP can then be regenerated at adequate levels to supply the body with energy.

ATP Production

About 95 percent of energy comes from our mitochondria and the balance from cellular glycolysis—the fermentation process that occurs in the cytoplasm of the cell and excludes mitochondrial participation. Normal mitochondria take up oxygen and nutrients, oxidize the nutrients, and store the energy created in ATP. Mitochondrial oxidative respiration—cellular respiration that requires oxygen—supplies the body with the energy it needs to function and relies on a steady source of oxygen from breathing.

ATP is generated by the cells in two pathways: fermentation (glycolysis) or respiration (cellular respiration requiring oxygen). Now you understand the importance of breathing—the oxygen taken in when you breathe powers the mitochondria for energy production. Glycolysis takes place in the cell cytosol, which isn't part of the mitochondria. The process takes one molecule of glucose and, through a series of ten steps, transforms it into two molecules of pyruvate. Once pyruvate is generated, the cell has a decision to make: It can take pyruvate into the mitochondria, where it will begin the respiratory energy cycle—the highly efficient process that employs oxygen to generate up to thirty molecules of ATP—or the cell can ferment pyruvate, an inefficient method of energy production that produces only two molecules of ATP and generates a byproduct, lactate, or lactic acid.

Excess lactic acid buildup in the body or bloodstream results in lactic acidosis, which is characterized by excessively low pH. Lactic acidosis can be caused by a chronic or acute medical condition, poisoning from medication, a prolonged lack of oxygen, low blood sugar, or excessive exercise. The symptoms include nausea, vomiting, abdominal pain, rapid breathing, weakness, and irregular heartbeat. The low pH isn't directly caused by lactate; it's created by excess hydrogens generated beyond the capacity of the mitochondria to reabsorb them.

There are three classes of macronutrients: carbohydrates, fats, and proteins. These are broken down during digestion into smaller, usable nutrients. Carbohydrates are broken down into glucose, and lipids (fats) are broken down into short-, medium-, and long-chain fatty acids. Proteins are digested by the proteases into simple amino acids and short chains of amino acids. These are the three fuels that cells use to produce ATP for the energy requirements of the approximately 100 trillion cells in the body.

To recap: ATP is a high-energy molecule called a nucleotide, which is easily converted into energy in most of the cell's parts. It's slightly unstable, with three phosphate groups attached to a ribose and an adenine. The energy of ATP is liberated by the enzyme ATPase, which cleaves one of the phosphate bonds on ATP, converting it to ADP.

Cellular Respiration

The primary biochemical pathway for ATP production in normal cells is through cellular respiration. Cellular respiration enzymatically converts glucose, fatty acids, and amino acids into ATP. Aerobic respiration requires oxygen to take pyruvate through the Krebs cycle and the oxidative phosphorylation steps that take place in the mitochondria. Anaerobic respiration—glycolysis—is a fermentation process that doesn't require oxygen and takes place inside the cell's interior, the cytoplasm. The glycolysis pathway metabolizes glucose into pyruvate, which generates two ATP molecules and two NADH molecules. NADH is a molecule that also enters the mitochondria and assists in electron chain transport to produce more ATP.

After generating ATP in glycolytic fermentation, the pyruvate is then enzymatically converted into lactic acid, a waste product. This glycolytic fermentation process allows cancer cells to survive, which will be discussed in greater detail in Chapter 10 on cancer. Cancer cells resort to an excessive reliance on the glycolysis pathway for energy production. This increases from the normal 5 percent seen in healthy cells to the vast majority of energy production occurring through the glycolysis pathway in cancer cells.

During aerobic respiration, pyruvate reacts with oxygen and can then be transferred into the mitochondria, where it forms acetyl coenzyme A. Pyruvate simply loses a carbon atom and bonds with coenzyme A to form acetyl coenzyme A. Acetyl coenzyme A then enters into the Krebs cycle, also called the citric acid cycle or the tricarboxylic cycle.

The Krebs Cycle

The Krebs cycle, named after Sir Hans Krebs, oxidizes pyruvate, fatty acids, and amino acids to carbon dioxide through a series of enzymatically controlled steps. The Krebs cycle takes place in the fluid-filled matrix of the mitochondria, the space between its two membranes.

The main purpose of the Krebs cycle is to produce eight high-energy electrons from these fuels, which are then transported by NADH and FADH2, two electron carriers, to the electron transport chain. The secondary purpose of the Krebs cycle is to produce precursors of several other key molecules (1, 2, 3).

The Electron Transport Chain

The next step in energy production is called the electron transport chain, the ETC. This process is also called oxidative phosphorylation, OXIPHOS, which occurs throughout the chain (4, 5, 6).

Electron transport takes place in the inner mitochondrial membrane, the cristae, in which a series of complexes numbered I, II, III, and IV is embedded. Each complex contains several different electron carriers. The energy released from the electron transport chain pumps protons between the mitochondrial membranes to create an energy gradient (6, 7, 8).

Complex I, known as NADH-coenzyme-Q reductase, accepts electrons from NADH to serve as the link between glycolysis, the Krebs cycle, fatty acid oxidation, and the electron transport chain (6, 7, 8).

Complex II is also known as succinate dehydrogenase. Complexes I and II both produce a compound called reduced coenzyme Q, which is used as the substrate for Complex III (7, 8).

Complex III, also known as coenzyme Q reductase, transfers

electrons from reduced coenzyme Q to produce cytochrome c, which is the substrate for Complex IV (6, 7, 8).

Complex IV, also referred to as cytochrome c reductase, produces water and two protons (7, 8).

Returning to our metaphor in Chapter 3 of mitochondrial respiration being analogous to the workings of a hydroelectric dam in which electrons run a series of pumps (the electron transport chain) that fill a reservoir of protons held back by a dam (the inner mitochondrial membrane) using an energy-producing turbine driven by the flow of electrons, Complex V represents the turbine. The only difference is that the energy flow is a chemical process. When a person exhales, moisture (water) and CO_2 (carbon dioxide) are released. These are the end products of the mitochondrial respiratory chain.

ATP synthase is the enzyme responsible for producing ATP from ADP plus phosphate through the final oxidative phosphorylation step. This process generates an additional thirty-four molecules of ATP (7, 8).

If that explanation was too technical for you, here's a short animation that illustrates how the process works:

https://www.youtube.com/watch?v=xbJ0nbzt5Kw&t=57s
Credit: http://vcell.ndsu.edu/animations

The Inefficiency of Electron Transport

Complex I is inefficient in transporting electrons in that it leaks some of the high-energy electrons that it generates. The electrons that are leaked react with oxygen to produce superoxide radicals, powerful free radicals that have a relatively long half-life. Superoxide radicals can travel through cell membranes because of their negative charge and can cause free-radical propagation in the cell membrane itself (7, 8, 9). Complex I leaks these electrons directly into the matrix, where the mitochondrial DNA is located. This leakage has dire health consequences, as we shall see later (7, 8, 9).

Complex III also has electron leakage, but not nearly as much the electron leakage from Complex I (8, 9, 10).

The electron leakage and superoxide generation in human mitochondria is approximately 1 percent at birth, and it increases to 2 to 3 percent in older mitochondria. This massive free-radical production causes approximately ten thousand mutations per cell per day and contributes significantly to the aging process (9, 10, 11, 12).

In the next chapter, we'll focus on the role of dysfunctional mitochondria in cardiovascular diseases.

References:

1. Lowenstein, J.M. *Methods in Enzymology, Volume 13: Citric Acid Cycle* (Boston: Academic Press, 1969).
2. Nelson, D. and Cox, M. "The Citric Acid Cycle," chap. 15, in *Lehninger Principles of Biochemistry*, 5th ed. (New York: W.H. Freeman and Company, 2008).
3. Ebenhöh, O. and Heinrich, R. "Evolutionary Optimization of Metabolic Pathways. Theoretical Reconstruction of the Stoichiometry of ATP and NADH Producing Systems." *Bulletin of Mathematical Biology* 63, no. 1 (January 2001): 21–55.
4. Voel, D. and Voel, J.G. *Biochemistry*, 3rd ed. (Hoboken, NJ: John Wiley and Sons, 2004). ISBN 978-0-671-58651-7.
5. White, D. *The Physiology and Biochemistry of Prokaryotes*, 2nd ed. (Oxford, UK: Oxford University Press, 2009). ISBN 976-0-19-512579.
6. Nichols, D.G. and Ferguson, S.J. *Bioenergetics 3* (Boston: Academic Press, 1999). ISBN 978-0-19-512579-5.
7. Nelson, D. and Cox, M.M. *Principles of Biochemistry*, 4th ed. (New York: W.H. Freeman and Company, 2005).
8. Lenaz, G and Genova, M.L. "Supramolecular Organization of the Mitochondrial Respiratory Chain: A New Challenge for the Mechanism and Control of Oxidative Phosphorylation. Mitochondrial Oxidative Phosphorylation." *Advances in Experimental Medicine and Biology* 748 (May 11, 2012): 107–44.
9. Droze, S. and Brandt, U. "Molecular Mechanisms of Superoxide Production by the Mitochondrial Respiratory Chain." *Advances in Experimental Medicine and Biology* 748 (May 11, 2012): 145–69.
10. Ames, B.N. "Delaying the Mitochondrial Decay of Aging—A Metabolic Tune-Up." *Alzheimer Disease and*

Associated Disorders 17 (April-June 2003): S54–7.

11. Liu, J., Killilea, D.W., and Ames, B.N. "Age-Associated Mitochondrial Oxidative Decay: Improvement of Carnitine Acetyltransferase Substrate-Binding Affinity and Activity in Brain by Feeding Old Rats Acetyl-L-Carnitine and/or R-Alpha-Lipoic Acid." *PNAS* 99, no. 4 (February 19, 2002): 1876–81.

12. Liu, J. and Ames, B.N. "Reducing Mitochondrial Decay with Mitochondrial nutrients to Delay and Treat Cognitive Dysfunction, Alzheimer's Disease and Parkinson's Disease." *Nutritional Neuroscience* 8, no. 2 (April 2005): 67–89.

Chapter 5

How Dysfunctional Mitochondria Affect Cardiac Health

"Optimum nutrition through metabolic cardiology appears to repair ailing hearts, and slows the progression of illness..."
—Stephen Sinatra

Introduction:

Your heart is an amazing organ that largely goes unnoticed throughout most of your life—unless or until a problem arises. If you take a moment and consider what's occurring in your heart—its constant contraction and expansion to move blood throughout the body—you immediately understand the significance of keeping your heart healthy.

Aging is known to be associated with alterations of various aspects of cell function. At the cardiac level, aging causes a decline in functional competence. Mitochondria play a central role in cardiac cell bioenergetics and are considered to be a locus of this decline. A crucial function in the regulation of mitochondrial energy metabolism is the transport of metabolites across the mitochondrial inner membrane.

Cardiovascular disease holds the No. 1 position in the list of top ten causes of death in America—above cancer, stroke, and diabetes. Cardiovascular-related disorders are a source of tremendous economic burden and quality-of-life deterioration for patients. Extensive biomedical research is devoted to elucidating the origins and mechanisms of cardiovascular disorders in an effort to target etiology and drive preventative measures. At the molecular level, mitochondria are central to energy production and the health of cells, including cardiac systems. Scientists have drawn ties between mitochondrial function and numerous cardiac irregularities known to affect heart disease and other detrimental injuries to the heart muscle.

This chapter will focus on the relationship of mitochondrial health to the well-being of the heart, drawing from examples of prevalent heart disorders. Although the dysfunction of mitochondria can result in serious injury to cardiac cells, there are numerous supplements and lifestyle regimens available to inhibit the aging process of the heart.

Heart Conditions Influenced by Mitochondrial Dysfunction

Mitochondrial Function and the Beating Heart

The heart is arguably the most important muscle in the human body, as proper functioning of this organ transfers roughly five to six quarts of blood to surrounding organs every minute. The heart functions with two main pumps that expand and contract to transfer blood through various arterial systems. Arteries supply the heart with oxygen-rich blood, and veins transfer poorly oxygenated blood from the heart to the lungs to pick up oxygen and send this freshly oxygenated blood out into the body. As with all biological processes—from cellular ones to those at the tissue level—this action requires energy. With each contraction and expansion of the heart to eject blood into the arteries, energy is required. The source of this energy is ATP, which is primarily supplied by the mitochondria. So, there's a direct link between healthy blood flow and the functionality of the mitochondria.

The mitochondria essential for cardiac energy production are located within cardiomyocytes (heart muscle cells). The cardiac system contains, on average, about five thousand mitochondria per cell—more than any other organ in the human body! In comparison, bicep muscle cells contain, on average, two hundred mitochondria per cell. This notably high concentration of energy-producing organelles in cardiac muscle reveals how hard the heart must work to supply the body with properly oxygenated blood—no small feat. The heart never turns off or slows down, so this exchange of energy, oxygen, and blood must persist twenty-four hours a day over the entire lifespan.

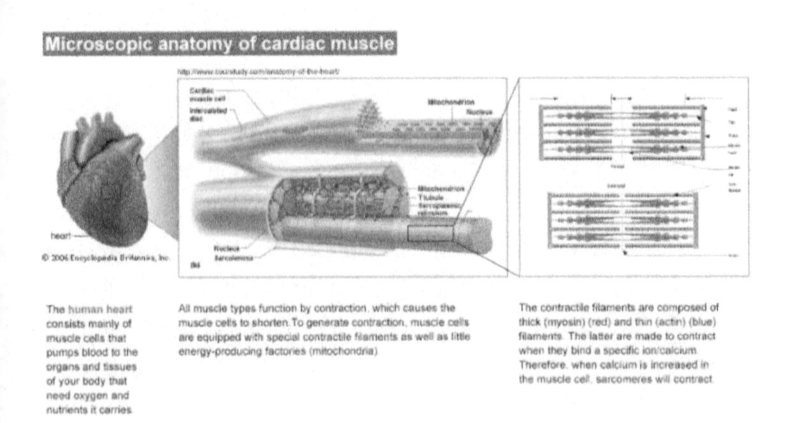

Figure 3. Anatomy of the Heart Muscle
Image courtesy of https://www.britannica.com/

Mitochondria must undergo oxidative phosphorylation to produce ATP to build up an energy pool for the heart to access, enabling it to pump blood effectively. Mitochondria are found in localized regions of cardiomyocytes to aid in energy production and metabolic support for the heart. With the potential of cardiac cells to accumulate calcium, mitochondria also have a hand in maintaining appropriate ion levels.

Cardiovascular Conditions Affected by Mitochondrial Energy Production

As the mitochondria proceed through the steps of oxidative phosphorylation, small amounts of oxygenated byproduct escape the electron transport chain (ETC). These byproducts of energy production are known as reactive oxygen species (ROS). It's been proposed that 0.2 to 2 percent of the molecular oxygen utilized in energy production is released as ROS, predominantly from Complexes I and III of the ETC (2).

Conditions in the heart can influence the levels of ROS produced in the mitochondria. Reperfusion refers to a process in which the body restores oxygen flow to tissues after a period of oxygen deprivation, such as what may occur during a heart attack when blood flow to cardiac tissue becomes temporarily blocked. Reperfusion results in heightened oxygen levels. During these times of cardiac rehabilitation, Complex I activity is suppressed, and the ETC produces more ROS (1). Higher levels of ROS are a potential danger to the human body because cells undergo oxidative stress if antioxidants don't intervene to counteract the toxic effects of ROS. Oxidative stress is a negative biological process that's known to be a precursor to several chronic diseases, including those afflicting the heart.

Studies in mice have shown that oxidative stress brought on by the suppression of antioxidants induces apoptosis in heart cells (4). The cascade of events that follows prolonged oxidative stress includes damage to the proteins, lipids, and DNA of the mitochondria, which disturbs mitochondrial function (2). Inhibited mitochondrial function leads to cellular death, which then propagates to the tissue level with the death and loss of cardiac cells following heart injury. Ultimately, the overproduction of ROS in mitochondria branches out to harm the body at the cellular level, disrupting energy-production mechanisms at the phenotypic level through exacerbated cardiac damage.

Aside from general energy production, mitochondria can also serve as indicators of cardiac injury or become facilitators of programmed cellular death in the case of irreversible injury to heart muscles. Mitochondria can detect detrimental damage to DNA and oxidative stress as signals to terminate the life of a cell through apoptosis. One mechanism that mitochondria undergo to induce cellular necrosis is to swell to abnormal size by opening the mitochondrial permeability transition pore and enabling the release of apoptosis-inducing factor and other related proteins (2). This mechanism causes inflammatory responses. Cellular death of both kinds has been observed in a multitude of heart conditions, including congestive heart failure and the damage and death of heart muscle tissue (2). By either mechanism, cardiac injury is characterized by a loss of cardiac cells through prolonged injury and cellular death.

Mitochondrial energy production and its role in cellular apoptosis have been linked to numerous cardiovascular injuries. This particular field of research provides growing evidence of the role of mitochondria in heart injury to direct the development of targeted treatment protocols and bolster basic understanding of chronic cardiac conditions. Dysfunction of mitochondrial energy production has been associated with heart conditions such as hypertension, congestive heart failure, angina, ischemia (blockage of blood flow), and diastolic dysfunction.

A common theme among these diverse cardiac injuries is an energy deficiency accompanied by inefficient energy production. The heart transitions from the effective method of oxidative phosphorylation to inefficient glycolysis in the absence of substantial oxygen levels. Glycolytic energy production is as detrimental as it is inefficient. The reliance on glycolysis to produce ATP leads to accumulation of the byproduct lactic acid. In anaerobic conditions, when the body is struggling to sustain oxygen levels, glucose is transformed into pyruvate. The body engineers a Band-Aid fix to maintain energy production by converting pyruvate into lactate, which can accelerate the breakdown of glucose. At high production rates, the lactate builds up in the body, causing acid levels to rise in muscles, decreasing the ability of muscles to contract. The heart combats this by expanding size over time, decreasing efficiency of blood intake and ejection and starving the heart of oxygenated blood.

Depleted cellular energy levels can cause a heart attack, while depleted oxygen levels induce the symptoms of a heart attack (3). Exacerbated heart injury is mostly indicated by lower cellular energy levels (up to 30 percent reduced) and purine leakage (4). The presence of excessive purines is observed in the buildup of uric acid, which can lead to gout. Thus, gout is a biological indicator of cardiovascular disease and low oxygen levels brought about by hypoxia or the clogging and swelling of the arteries in atherosclerosis.

Cardiometabolic Syndrome

Mitochondrial capacity and insulin resistance serve as key indicators of cardiometabolic syndrome, which will be discussed at length in the next chapter. In brief, cardiometabolic syndrome is characterized by the accumulation of fat in the abdominal area, insulin resistance, and glucose intolerance and is often a precursor to stroke, heart attack, and diabetes. Insulin resistance is a hallmark symptom of several cardiometabolic disorders, including hypertension, congestive heart failure, coronary artery disease, and stroke (7).

Free fatty acids circulating within the bloodstream can negatively affect mitochondrial oxidative ability and cardiac output in terms of energy production: they also reinforce insulin sensitivity (7). The oxidative capacity of the mitochondria provides a direct indication of the stability of the cardiac system. The heart is increasingly susceptible to oxidative stress due to rather low antioxidant capacity (7). It's the responsibility of mitochondria to reduce production of ROS, or we can utilize supplements as antioxidants to remove these harmful species from circulation. Prolonged oxidative stress has the power to alter systolic and diastolic function in the heart as well as impede insulin-signaling pathways (7). A sufficient number and quality of functioning mitochondria are imperative to prevent the oxidative stress that reinforces many adverse processes in the cardiometabolic system, including insulin resistance, restricted blood flow, and fatty-acid accumulation.

The influence of mitochondrial energy levels on heart function directly affects diastolic function and the heart's ability to efficiently intake deoxygenated blood and eject oxygenated blood to surrounding muscles and organs. Low ATP levels affect calcium ion channels in the heart. These portals for calcium entry and exit are disabled in low-energy environments so that ions can't escape heart muscles.

Calcium (Ca) is one of the most important ions in the human body. In the context of heart muscles, calcium accumulates in the heart upon contraction, with subsequent ejection from the heart by ATP to allow relaxation of heart muscle (5). If the heart can't expel Ca ions from cardiac muscle, it can't properly relax and fill with blood, a phenomenon known as diastolic dysfunction (3). The electrochemical gradient involving ions such as potassium (K), Ca, and sodium (Na) is essential for a healthy heart (3). This is a highly regulated process just like energy production. A person's ability to endure stress on the heart is largely dependent on the quality of energy storages, the ability to transfer ions across membranes, and the level of mitochondrial dysfunction.

Cardiovascular Disease and Atherosclerosis

Cardiovascular disease and its clinical sequelae, atherosclerosis, are the leading causes of morbidity and mortality in the Western world. An elevated level of LDL is associated with increased risk of coronary artery disease (8). Oxidative modification of LDL and its transport into the subendothelial space of the arterial wall at the sites of endothelial damage is considered an initiating event for atherosclerosis (9).

The heart's coronary artery system provides oxygen and nutrients for the heart's mitochondria and removes the byproducts of carbon and water resulting from mitochondrial energy production. The coronary artery system overlays the mitochondria-dense muscles comprising the heart (see Figure 4). The three major coronary arteries—left anterior descending (LAD), circumflex, and right coronary artery (RCA)—and their respective branches each supply a designated portion of the heart with blood. The LAD supplies blood to the front (anterior) portion of the heart and septum (partition that separates the left ventricle and right ventricle); the circumflex supplies the back (posterior) portion of the left ventricle; and the RCA supplies the bottom (inferior) portion of the left ventricle and the right ventricle.

Coronary arteries have muscle fibers within their walls. When the muscles within the artery walls contract, blood flow within the artery is reduced; relaxing the muscle increases flow. In this way, the coronary arteries can regulate blood flow to different portions of the heart.

Previous chapters have identified that increased mitochondrial oxidative damage is a major feature of most age-related human diseases, including atherosclerosis.

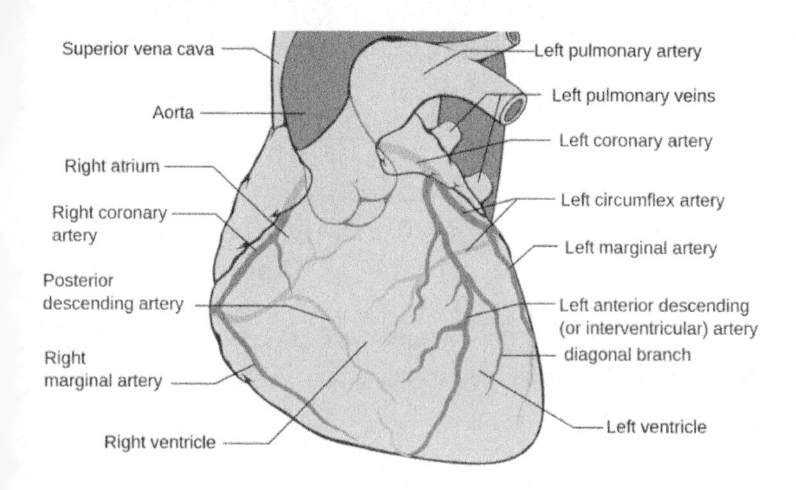

Figure 4. Anatomy of the Heart
Courtesy: https://en.wikipedia.org/wiki/Coronary_circulation

Under normal physiological conditions, ROS are produced by the oxidative phosphorylation (OXIPHOS) pathway involved in energy production in mitochondria and are removed by antioxidants. In chronic conditions, ROS generation in mitochondria is dysregulated by a number of factors, such as lack of oxygen (hypoxia), caloric excess, lack of exercise to drive mitochondrial biogenesis, and lack of essential nutrients to drive the electron transport chain.

In addition to the mitochondria being the major site of ROS production, the mitochondria are also the first to be compromised by prolonged exposure to ROS—specifically, the cristae (inner membrane of the mitochondria) and the DNA located within the mitochondria (mtDNA). In fact, heart tissue from patients with coronary artery disease demonstrates eight to two thousand times more mtDNA deletions than cardiac tissue from age-matched controls (10).

In Chapter 3, "Theory of Aging," we discussed a process that described how dysfunctional mitochondria drive aging, with the end product being oxidized LDL distributed both within and outside of cells.

Oxidative modification of LDL—and its transport into the subendothelial space of the arterial wall at the sites of endothelial damage—is considered an initiating event for atherosclerosis (9, 12).

Now, let's summarize the initiating events for the pathogenesis of cardiovascular disease, with an emphasis on atherosclerosis plaque formation.

1. ROS is produced by the mitochondria under normal conditions. In chronic conditions (hypoxia, caloric excess, hypertension, lack of exercise), ROS production is increased.
2. When ROS comes in contact with lipids, the lipids become oxidized.
3. The coronary artery system overlays the cardiac muscles, which have the highest concentration of mitochondria and ROS output. Additionally, the endothelium (the inner lining of the artery) contains mitochondria.
4. LDL is used to deliver cholesterol to the cells for membrane repair. If the LDL is oxidized by ROS, it becomes trapped in the three-dimensional matrix of fibers and fibrils secreted by cells in artery walls.
5. Once the oxidized LDL is trapped in the artery walls, the body initiates inflammatory responses.

Specific Effects of Low Oxygen, D-ribose, CoQ10, and L-carnitine on the Heart

Extensive research has deduced a trifecta of compounds that are imperative for the rehabilitation of an oxygen-starved heart: D-ribose, coenzyme Q10 (CoQ10), and L-carnitine. If damaged heart muscle is deprived of these nutrients, the tissues will continue to deteriorate. These key compounds, the stability of mitochondrial energy production, and oxygen levels in the heart interact to determine the prognosis of a stressed and damaged cardiac system.

D-Ribose

It's been suggested that deoxy-ribose (D-ribose) is the most beneficial supplement to heart health of the three compounds mentioned above (2). D-ribose is a sugar molecule present in many biological pathways—predominantly genetic building systems and metabolic pathways. D-ribose can be consumed through the diet, but the amount is often insufficient to truly remediate heart damage (5). The most bioavailable source of D-ribose derives from the pentose phosphate pathway, where it can be further absorbed directly into the blood (4). D-ribose is a primary output of the pentose phosphate pathway to create nucleotides for DNA molecules and promote cellular metabolism (3). Aside from its place in human genetics, D-ribose is a five-carbon monosaccharide that's been shown to be necessary for maintaining a healthy heart.

D-ribose supports the heart's ability to pump blood by bolstering energy transformation. This therapeutic compound is unique to energy metabolism, because no other sugar exists that can keep cells alive by preserving energy compounds and regulating energy-making pathways (3). One of the most important remedies that D-ribose can offer an injured heart is replenishment of the purine pool.

The purine pool is a collection of molecules that's constantly expended and restored to drive the reaction of the combination of two ADP molecules to form ATP and AMP (adenosine monophosphate). AMP is discarded as ATP is used for the cell's high-energy source. The degradation of AMP leaves behind the compounds adenine, inosine, and hypoxanthine, which, given enough time, can amalgamate to form ribose (5). Purines are lost from this conversion, and it's essential to rebuild the purine pool, especially in an oxygen-poor environment, because this deprivation ultimately will lead to death (3). If a heart undergoes an ischemic episode (where blood flow is blocked), it can take nearly ten days to refuel the purine pool. D-ribose trumps this timeline by rebuilding the pool in a short one to two days (3).

Since D-ribose is produced from the pentose phosphate pathway, it's essential that the primary substrate of this pathway, glucose, is adequately stored. D-ribose availability is directly linked to glucose availability. Mitochondrial dysfunction complicates this necessity because, when the heart is starved of oxygen, energy production transfers from efficient oxidative phosphorylation to glycolysis (5). Out of obvious need to keep the heart beating with properly oxygenated blood, the mitochondria shift to utilizing all available glucose for the process of glycolysis rather than building D-ribose.

Through these incredibly important mechanisms, D-ribose is able to replenish energy levels that drastically decline during ischemia. Supplementation has the potential to revitalize dormant portions of the heart after cardiac injury, surgery, and heart failure, and it also repairs muscle (4). Regarding intense physical activity, athletes that train at high aerobic levels can benefit from D-ribose supplementation, which also reduces the risk of a slowing heart rate (bradycardia) that often accompanies exercise-induced stresses to the heart (4). The case has certainly been made for the tight integration of D-ribose with heart functioning and maintaining a properly energized cardiac system.

Coenzyme Q10

Though D-ribose was proposed to be the most vital compound for heart health, coenzyme Q10 (CoQ10) has been deemed the most vital nutrient to sustain mitochondrial health. CoQ10's role in the human body is extensive; it's present in nearly every cell of the body and is considered invaluable for life itself.

It's necessary to supplement the body with CoQ10, because the aging process naturally leads to depleted production of this nutrient. Aside from the effects of aging, poor nutrition can also expend CoQ10 levels in the body. Specific deficiencies in vitamins B6, B12, and C and folic acid can induce lowered levels of CoQ10 as a result of malnutrition (2). Additionally, it's been proposed that statin use is linked to lower CoQ10 levels in the body due to inhibition of the enzyme HMG-CoA reductase (2). HMG-CoA reductase is an important enzyme in the mevalonate pathway, a biological process that produces cholesterol and quinones, which are molecules used to accept and transfer electrons during the electron transport chain that produces ATP.

CoQ10 plays a starring role in the mitochondrial electron transport chain by serving as a redox agent to transfer electrons among the three main complexes. CoQ10 moves along the ETC by first accepting an electron from Complex I or II and, with the acceptance of an extra electron, CoQ10 transforms into a reduced form. CoQ10 then carries this electron over to Complex III, transforming into an oxidized form by donating the electron from Complex I/II to Complex III (5). Aside from shuttling electrons through the ETC, CoQ10 also serves as a protective agent for mitochondria by stepping up as an antioxidant when free radicals escape from Complex I. CoQ10 recycles these harmful oxygen species by returning them to the ETC (5). By acting as an antioxidant, CoQ10 prevents potential damage to the mitochondrial DNA as well as surrounding lipids and proteins (5).

The applications of CoQ10 to sustain health during aging range from heart benefits to neural protection. In terms of the brain, CoQ10 has been used in studies focused on Huntington's disease, Parkinson's disease, and Lou Gehrig's disease (amyotrophic lateral sclerosis, or ALS) (5). CoQ10 prevents nerve cells in the brain from damage caused by excitotoxicity, an adverse effect of severe brain injury such as stroke or neurodegenerative disorders. CoQ10 protects nerve cells from excitotoxicity by allowing energy production to access these neural cells. CoQ10 intervenes to allow pore transfer into the membrane and promote anti-inflammatory effects (5). All these neural disorders have the power to drastically alter a human's life and, in many cases, accelerate mortality.

CoQ10 not only provides essential neuroprotective effects, but it also has a large effect on cardiac sustainability. Numerous cardiac injuries are worsened by the effects of hypertension (high blood pressure), indicating restricted blood and oxygen flow to the heart. CoQ10 has the ability to significantly lower blood pressure levels by diminishing the amount of ROS in the blood. The amount of free radicals is heightened during periods of cardiac stress because the electron transport chain backs up and spills free radicals in times of low oxygen availability (5). Such adverse effects add up when the heart is starved of blood for prolonged periods of time, which occurs in such conditions as weakened heart muscle, coronary heart failure, and hypertension (5). In addition to the benefits of lowering blood pressure, CoQ10 also prevents damage caused by plaque formation in the arteries. By reducing the stickiness and size of platelets in the blood, arteries are less likely to become blocked (3). Blockages caused by oxidized LDL are capable of solidifying to a more advanced condition called atherosclerosis, where arteries begin to narrow and restrict blood flow to the heart through the aorta. CoQ10 addresses these risks by preventing the oxidation of LDL, which helps to promote energy production in the electron transport chain to supply the heart with adequate ATP levels and allow for its relaxation after contraction. In prior studies, CoQ10 has shown beneficial effects on hypertension, congestive heart failure, coronary artery disease, irregular heartbeats, and chest pain (3).

L-Carnitine

The final of the three compounds used for improving heart mitochondrial health is L-carnitine, which is a product of lysine and methionine and occurs naturally in the human body (4). Food sources of L-carnitine include meat and dairy products, which means vegans and vegetarians are often deficient in this nutrient (5).

One of the main causes of deficiencies in L-carnitine is the aging process. As we grow older, L-carnitine levels naturally decline, implying adverse effects on the heart and the development of other medical conditions. In an almost circular fashion, deficiencies in L-carnitine are brought about by aging and therefore speed up the aging process, further reducing L-carnitine levels (5). Aside from its effects on the cardiovascular system, low supplies of this compound have been linked to liver disease, genetic abnormalities, and neural disorders associated with memory issues, such as dementia and Alzheimer's disease (5). When maintained healthy levels, L-carnitine can serve as a protective agent to prevent development of these neurodegenerative disorders.

In conditions of hypoxia, energy production shifts to an inefficient anaerobic metabolism, which produces lactate as a secondary product. As mentioned previously, lactate buildup in the muscles of the heart can induce angina, or chest pain, as a marker of restricted blood flow to the heart. In brief, lactic acid causes damage to tissues and muscles. Angina is a common symptom of coronary artery disease. L-carnitine steps in to promote aerobic processes to decrease the production of lactate via anaerobic metabolism (5). Carnitine functions within the cells to transport fats and triglycerides, specifically to mitochondrial Complexes II–V (5). Once these fats and triglycerides enter the mitochondria, they undergo beta-oxidation to be converted into cellular energy as ATP (5).

It's necessary to transport fats—the desired energy source—into cells because they cannot enter without an attached acyl group (5). Furthermore, once inside the mitochondria, L-carnitine wraps around the fats to remove the attached acyl group so these fats can be used in beta oxidation.

Therefore, there are two main derivatives of L-carnitine—one with an attached acyl group and one without an acyl group. The acyl group is essentially a tag for entry into the mitochondria and a blockade for metabolism of fatty-acid chains for energy production. L-carnitine is so pivotal to the oxidation of fats for cellular energy that the rates of these metabolic processes are contingent on the bioavailability of L-carnitine, and levels of L-carnitine in the cell can be indicative of the health of the functioning mitochondria. Through the prevention of lactate production in the cells, L-carnitine helps sustain oxygenated blood flow via vasodilation to the heart and prevents further cardiac injury (3).

The Effects of Calorie Restriction on Heart Mitochondria

Caloric restriction is a tactic most often employed to lose weight and prevent obesity. In partnership with these outward benefits, calorie restriction has also shown potential to improve cardiometabolic syndrome and diabetes as well as to delay mortality (7). The specific mechanisms of improvement are related to the metabolism of glucose and lipids. On the genetic level, several studies have shown the capability of caloric restriction to reverse the slowing of metabolism that aligns with the aging process (7). Limiting caloric intake achieves these goals through the production of more mitochondria within the cell, improved consumption of oxygen to support aerobic energy metabolism, and heightened energy output through increased concentrations of ATP (7).

In several studies that have attempted to explain the mechanism by which calorie restriction leads to an increased production of mitochondria within the cell, this process was linked to increased SIRT (also known as sirtuin-1 and NAD-dependent deacetylase sirtuin-1) protein levels and NO (nitric oxide) production in addition to reduced production of ROS (7). Expanding the population of mitochondria within the cell improves lipid metabolism, insulin sensitivity, and glucose tolerance. It also can strengthen the cardiovascular system and prevent metabolic syndrome.

The Effects of Exercise on Heart Mitochondria

Exercise has long been associated with overall health benefits, including the prevention of chronic diseases such as cancer, diabetes, cardiovascular-related conditions— and delayed mortality. Research on the beneficial effects of exercise ranges from large-scale, epidemiological studies of trends within the population to biochemical, in vitro studies to understand the mechanism of physical activity on health. At the cellular level, exercise has been shown to improve the production of mitochondria in addition to enhancing mitochondrial size and promoting glucose metabolism through oxidative activity (7). In fact, mitochondrial biogenesis resulting from increased levels of physical activity has been shown to be one of the main benefits of exercise on energy production. With the formation of new mitochondria, the cell is able to increase energy output.

In diabetic patients, exercise improves insulin sensitivity and glucose tolerance (7). Decreased mitochondrial function and restricted aerobic energy metabolism are strong risk factors for the development of cardiometabolic syndrome (7). Cardiometabolic syndrome greatly increases the likelihood of developing type 2 diabetes or atherosclerotic vascular disease, or experiencing a stroke or heart attack (8). Insulin resistance and glucose intolerance are hallmark characteristics of cardiometabolic syndrome (8). As humans age, their production of mitochondria decreases, so exercise serves as a natural remedy to combat depleted production of these necessary structures for energy production.

Basic Protocol for Cardiac Rehabilitation

Rehabilitation of heart muscle after cardiac injury can be promoted with the supplementation of several key nutrients. The primary goal of cardiac rehabilitation therapy is to sustain the heart in pumping adequately oxygenated blood. This is achieved through efficient energy metabolism at increased rates of production (3). A widely accepted treatment regimen for damaged heart muscles involves supplementation with CoQ10, L-carnitine, and D-ribose, as described above. Though these compounds are central to the discussion of cardiac supplement therapy, other pharmacological agents and natural supplements are available on the market to combat further cardiac injury after an episode of cardiac damage.

THERAPEUTIC INTERVENTION

Pharmacological Intervention

One option to restabilize an injured cardiovascular system is through pharmacological agents designed to increase blood flow to the heart and improve cardiac output. At the cellular level, these agents often operate to stimulate mitochondrial biogenesis. Thiazolidinediones such as AvandiaTM or ActosTM mediate insulin sensitivity through increased production of mitochondria (7). By mediating insulin sensitivity, this class of drugs may help prevent cardiometabolic syndrome. Metformin is also used to improve insulin sensitivity while increasing the amount of mitochondria per cell and reducing the number of free-radical oxygenated species that can induce oxidative stress (7). Though these are only a few examples of pharmaceutical options, ample research has been devoted to develop pharmacological targets to improve mitochondrial function and reduce angiotensin levels in the heart.

Though widely used to restore heart function after cardiac energy, there are several shortcomings of pharmacological approaches to cardiac rehabilitation. Some of the most commonly prescribed medications used to mediate heart health include beta-blockers, aspirin, nitrates, ACE inhibitors, and calcium channel blockers (3). The purpose of these pharmacological agents is to increase the blood supply to the heart after restriction caused by prior injury. For example, inotropic drugs are used as pharmaceutical agents to influence the speed of muscle contraction. Though these agents are used to improve contraction in the heart, this remedy is ineffective because, although these drugs elicit a strengthened heartbeat, the body can't support this improved cardiac activity in terms of energy needs (3).

In another example, statins are commonly used to lower cholesterol, which, if left uncontrolled, can induce a myriad of negative health effects. In sequence with controlling cholesterol levels, statins deplete CoQ10 levels in the cells by interfering with HMG-CoA reductase, a critical enzyme in the metabolic pathway (mevalonate pathway) that produces cholesterol (2). While HMG-CoA reductase is a logical target because it's involved in cholesterol production, CoQ10 is also a product of this enzyme. Furthermore, beta-blockers and diabetic medications depress CoQ10 levels, which are already declining in an aging individual (5).

The primary reason these medications aren't effective is they don't address the issue of energy deprivation, which is one of the hallmark characteristics of cardiac injury and cardiovascular disease. Although they may provide a temporary fix in supplying blood to the heart, they don't prevent further injury, and, instead, result in a heart with improved responses but not enough energy to sustain such actions.

Nutrition for Mitochondria
(Mostly for Cardiovascular Disease but for Cognition and Aging as Well)

Several nutrients have been studied in relation to heart benefits and are available as supplements to replenish an oxygen-starved heart or to prevent further cardiac injury.

Magnesium

In the human body, magnesium is a vastly utilized ion in numerous biochemical reactions that take place in many organ systems—over three hundred reactions, in fact (5). Magnesium is known to share a role in the production of ATP at the cellular level. Magnesium and another highly prevalent ion, calcium, are interrelated to an amazing degree in biological processes. Magnesium can be used as a sort of natural calcium channel blocker to promote relaxation of the heart muscle (3). If calcium entered the heart muscle cells and remained there without the addition of magnesium, the heart muscle would remain tensed and contracted (5). Magnesium supports ATP and other heart enzymes to promote relaxation after contraction. Adverse effects of prolonged contraction include tightening of the blood vessels around the heart, known as vasoconstriction (5). A side effect to narrowing of the blood vessels is suppressed oxygen supply to the heart, which produces high blood pressure and diminishes the ability of mitochondria to produce energy. Increased blood pressure and decreased energy production in the heart for extended periods of time have been shown to induce a slew of injuries, including coronary heart failure, angina, ischemic heart disease, coronary artery disease, death of cardiac muscle tissue, insulin resistance leading to diabetes, and even asthma (5). The necessity of sufficient magnesium in the diet is profound for the health of heart muscles and the stability of energy production.

Alpha-Lipoic Acid

Present within mitochondria as an antioxidant, alpha-lipoic acid (ALA) serves the mitochondria by holding NAD/NADH levels constant to ensure stable energy storage. It's important for the cell to maintain high levels of NAD+ and NADH because these molecules are used as oxidizing and reducing agents to support oxidative phosphorylation through the electron transport chain by shuttling electrons. The reduced form gains an electron to become NADH, and the oxidized form donates an electron to become NAD+. Excessive glucose levels in the cell prevent the conversion of NADH to NAD (4). This conversion backup produces several downstream effects, such as increased free-radical damage, electron accumulation at Complex I of the electron transport chain, and depletion of iron storages (5). ALA is also valued as an antioxidant agent to suppress the number of ROS within the cell that leads to oxidative stress and insulin sensitivity (3). Supplementation with ALA contributes anti-aging effects to slow cognitive decline, improve cardiac output, and maintain physical activity capabilities.

B Vitamins

Members of the family of B vitamins play individual as well as integrated roles in mitochondrial energy output to protect the aging heart and cardiovascular system. Vitamin B1, found naturally in wheat, yeast, and legumes, participates in the carbohydrate metabolic process by facilitating the conversion of pyruvate to acetyl-CoA for the TCA cycle, which works to produce ATP and provide the body with adequate high-energy compounds. Vitamin B1 exhibits benefits for heart sustenance as well as neural protection (5). Vitamin B2, another member of the family of B vitamins, is available in nuts, yeast, whole grains, and other food sources.

Vitamin B2 is found directly in the electron transport chain at Complex I and II. It assists mitochondrial cofactors in electron movement from the ATP products of the TCA cycle to Complex II (5). This mechanism is beneficial in cases of a shortage of Complex I or impaired functionality (5). Next up is Vitamin B3, which is found in meat, fish, and dairy products. Vitamin B3 is related to NAD/NADH levels, as it serves as a precursor for these high-energy compounds (5).

In maintaining stable levels of NAD/NADH, vitamin B3 is valuable for its anti-inflammatory and antioxidant capabilities in relation to cancer, diabetes, and neurodegenerative disorders (5). Vitamin B5 is present predominantly in vegetables, meats, and whole grains. Vitamin B5 assists acetyl-CoA to transport pyruvate into the TCA cycle and stimulate production of ATP.

This B vitamin also plays a role in metabolizing carbohydrates and fats. Both of these mechanisms provide the cardiac system with stable supplies of energy to combat and prevent heart injury. The last B vitamin to note is vitamin B12. Supplementation of B12 is essential for the body because it's not naturally synthesized within humans. Dietary sources include mostly animal products, such as meats, fish, and dairy. Vitamin B12 has a large role in cell survival and energy production (5). Low levels of vitamin B12 can increase the odds of suffering from heart disorders, such as angina, stroke, and heart attack (5).

Pyruvate

Pyruvate and its connection to mitochondria provide leverage for energy metabolism benefits and remedies for an injured heart. Pyruvate is an essential intermediary in many metabolic pathways, substantially influencing energy production and mitochondrial health.

Pyruvate can be synthesized from glucose in glycolytic energy production. Depending on the needs of the cellular environment, pyruvate can be converted back into glucose through a process known as gluconeogenesis or further metabolized to produce fatty acids or lactate. With such a strong presence in the energy production process, pyruvate plays a starring role in energy metabolism within the heart (6).

The next chapter focuses on the relationship of mitochondrial dysfunction and the development of cardiometabolic syndrome.

References:

1. Sinatra, S.T. The *Sinatra Solution: Metabolic Cardiology* (North Bergen, NJ: Basic Health Publications, 2005), Kindle.
2. Gustafsson, Å.B. and Gottlieb, R.A. "Heart Mitochondria: Gates of Life and Death." *Cardiovascular Research* 77, no. 2 (2008): 334–43. doi:10.1093/cvr/cvm005
3. Sinatra, S.T. *The Sinatra Solution: Metabolic Cardiology*. (North Bergen, NJ: BasicHealth Publications, 2005), print edition.
4. Siwik, D.A., Tzortzis, J.D., Pimental, D.R., Chang, D.L., Pagano, P.J., Singh, K., et al. "Inhibition of Copper-Zinc Superoxide Dismutase Induces Cell Growth, Hypertrophic Phenotype, and Apoptosis in Neonatal Rat Cardiac Myocytes in Vitro." *Circulation Research* 85 (1999): 147–53.
5. Nd, L.K. *Life—The Epic Story of Our Mitochondria: How the Original Probiotic Dictates Your Health, Illness, Ageing, and Even Life Itself* (Victoria, BC: FriesenPress, 2014).
6. Paradies, G. "The Effect of Aging and Acetyl-L-Carnitine on the Pyruvate Transport and Oxidation in Rat Heart Mitochondria." *FEBS Letters* 454, no. 3 (1999): 207–9. doi:10.1016/S0014-7935 (99)00809-1
7. Kim, J., Wei, Y., and Sowers, J.R. "The Role of Mitochondrial Dysfunction in Insulin Resistance." *Circulation Research* 102 (2008): 401–14. doi:10.1161/CIRCRESAHA.107.165472
8. Srivastava A.K. "Challenges in the Treatment of Cardiometabolic Syndrome." *Indian Journal of Pharmacology* 44, no. 2 (2012): 155–6. doi:10.4103/0253-7613.93579.
9. Steinberg, D., Parthasarathy, S., Carew, T.E., Khoo, J.C., and Witztum, J.L. "Beyond Cholesterol.

Modifications of Low-Density Lipoprotein That Increase Its Atherogenicity." *New England Journal of Medicine* 320 (1989): 915–24.

10. Witztum, J.L. and Steinberg, D. "Role of Oxidized Low-Density Lipoprotein in Atherogenesis." *The Journal of Clinical Investigation* 88 (1991): 1785–92.

11. Navab, M., Berliner, J.A., Watson, A.D., Hama, S.Y., Territo, M.C., Lusis, A.J., Shih, D.M., Van Lenten, B.J., Frank, J.S., Demer, L.L., Edwards, P.A., and Fogelman, A.M. "The Yin and Yang of Oxidation in the Development of the Fatty Streak. A Review Based on the 1994 George Lyman Duff Memorial Lecture." *Arteriosclerosis, Thrombosis, and Vascular Biology* 16 (1996): 831–42.

Chapter 6

Mitochondrial Dysfunction and the Development of Cardiometabolic Syndrome

Introduction

Cardiometabolic syndrome is a precursor of countless chronic diseases: cardiovascular disease, atherosclerotic vascular disease, diabetes, hypertension, and many others. As a determining factor for some of the world's most detrimental chronic diseases, it's essential to understand the effect of cardiometabolic syndrome on morbidity and mortality in cardiovascular-related disorders.

Under the umbrella of cardiometabolic syndrome are several symptoms, including insulin resistance, glucose intolerance, hypertension, abdominal obesity, and high levels of cholesterol in the bloodstream (1). Properly functioning insulin is an absolute necessity to maintain healthy glucose production and processing. Glucose metabolism takes place primarily in skeletal muscles, the liver, and the pancreas (2). Glucose metabolism and ATP production are linked because, when glucose levels rise, ATP production is increased. Pancreatic beta cells use ATP to produce insulin and reduce glucose levels in the body. It's been reported that insulin resistance is present in all patients with type 2 diabetes, sometimes dating back decades before the onset of the disease (3).

Approximately 7 percent of the U.S. population has diabetes, and the prevalence continues to increase. Type 2 diabetes comprises 90 to 95 percent of all diabetic cases in the world. Type 2 diabetes is a serious cause for concern in Western society, and it's projected that more than 250 million people annually will be diagnosed with the disease by 2020, causing a tremendous health and financial burden worldwide (3). The incidence of type 2 diabetes is differentially higher in African American and Hispanic populations (4). Thus, diabetes and the related problems of obesity and vascular disease represent major global health care issues (4). Telltale symptoms of diabetes include increased urine production, excessive thirst, and slow healing of injuries and wounds (5).

One study on diabetic patients found that proper glycogen production in the muscles is essential to maintain proper insulin function, and insulin resistance in type 2 diabetic patients is caused by defective glycogen synthesis (6). Glycogen is a long-term energy storage form of glucose, commonly kept in the liver or muscles and regulated by insulin. Glycogen levels can be either inflated or depleted in diabetic patients, but the proper synthesis of glycogen often goes hand in hand with normal glucose metabolism (7). Insulin resistance can inhibit glycogen synthesis through suppression of glucose transport within the cell or glucose phosphorylation, key steps involved in the synthetic pathway of glycogen (3). The negative effects of insulin resistance are hallmark indicators of the risk for type 2 diabetes, which may even be witnessed in the children of diabetic patients, whose risk for later development of type 2 diabetes is nearly 40 percent higher (3). The magnitude of influence that blood sugar stabilization has on chronic conditions such as type 2 diabetes sheds light on the importance of understanding the cellular mechanisms involved in maintaining healthy blood sugar levels.

Insulin Resistance and Glucose Levels in Type 2 Diabetes

A cornerstone to the discussion of type 2 diabetes and blood sugar control is impaired insulin sensitivity and heightened glucose levels in the blood. When a patient is insulin-sensitive, his or her body is resistant to the key mechanisms that insulin enacts to maintain proper blood sugar control. These processes primarily consist of the uptake of glucose by skeletal muscles and suppressing glucose production in the liver when blood sugar levels are adequately high (7). Defective receptors on the insulin signaling pathway also lead to the development of insulin resistance (8). In short, insulin resistance from abnormal insulin production desensitizes tissues to the metabolic effects of insulin (5). Insulin resistance develops through abnormal glucose processing and lipid metabolism, which directly affect the signaling capabilities of insulin within the body (8). At the cellular level, blood sugar control is impaired in those suffering from type 2 diabetes because their cells experience low rates of glycogen synthesis within the muscles, which bolsters insulin resistance in the cells (3). As a result, abnormally high glucose levels are seen in the blood, and blood sugar regulation is impaired. The interplay of insulin resistance and glucose transport and phosphorylation (the addition of a phosphate group to ADP to form ATP) is the most pertinent relationship in understanding blood sugar control in diabetic patients.

Along with inflated glucose levels in the blood, excessive amounts of fatty-acid metabolites such as triglycerides and fatty acids in cells is a hallmark symptom of insulin resistance and the diabetic condition. In an insulin-resistant individual, skeletal muscles and the liver struggle to oxidize fatty acids, leading to abnormal lipid levels in the cells. A person may experience elevated lipid levels in the blood as a result of obesity resulting from excessive caloric intake, stress-induced states, or disorders in the metabolism of fatty acids (8). Obese individuals have adipose tissue that's overwhelmed by excess triglycerides, leading to disorders such as dyslipidemia, which is characterized by abnormal lipid levels in the blood (8). The accumulation of fatty acids primarily in the skeletal muscles and the liver can cause defective glucose processing through impaired insulin signaling and glucose-activated pathways to promote the synthesis of insulin (3). Additionally, high levels of fatty acids in the cell have been shown to impede the regulation of genetic processes responsible for oxidative phosphorylation and the production of new mitochondria (13). Excessive intracellular lipid concentrations are also witnessed in cardiovascular tissues, extending this abnormality to cardiovascular diseases such as hypertension, stroke, coronary artery disease, and congestive heart failure (8). The issues of insulin resistance and blood sugar regulation aren't unique to diabetes, but they represent a confounding symptom of cardiometabolic syndrome. Ultimately, increased fatty-acid concentrations elevate the effects of insulin resistance by impeding glycogen synthesis in the muscles and surrounding tissues, linking these biological processes together to exacerbate the diabetic condition.

Mitochondria and Blood Sugar Regulation

Once we understand the mechanisms of insulin resistance and glucose accumulation in the cells of patients with unstable blood sugar regulation, it's important to establish a link between these mechanisms and the various processes that contribute to mitochondrial dysfunction. The mechanisms at the forefront of this discussion include an influx in ROS, oxidation of membrane lipids, excess caloric intake, lack of sustained physical activity, and the natural process of mitochondrial degradation from aging. The mitochondria are the energy source of the cell and, over time, the concentrations of mitochondria diminish and their functionality declines. As mentioned in previous chapters, defective mitochondrial processing results in an imbalance between energy input and energy output such that there's deficient energy production in the cell (8). Abnormalities in the number, size, or function of mitochondria can severely affect energy production in the cell, which extends to further development of oxidative stress and chronic disease.

Reactive Oxygen Species Production

There are several risk factors that promote mitochondrial irregularities aside from the natural process of aging. Subunits that make up the mitochondria are coded by genes in the nucleus and the mitochondria (7). Mutations in these coding schemes can cause irregularities in mitochondrial function. Errors in the genetic code for mitochondria can be induced by oxidative stress, another factor that contributes to dysfunction in cellular energy production. The genetic building blocks for mitochondria are located in close proximity to where reactive oxygen species (ROS) are produced (7). Endothelial cells in close proximity to high glucose levels are susceptible to perturbations by ROS, which has been proposed to damage the vascular system and its function in patients with diabetes (4). ROS production accelerates when too many electrons are supplied to the electron transport chain and they're displaced to oxygen, forming highly unstable oxygenated species that can cause damage to the cell in the form of oxidative stress (7). When there are abnormally high ROS levels in the cell, the demand for ATP is low, there's excess calorie intake, accelerated aging, and low levels of physical activity (7). This could be why, as individuals age, they accumulate more fat around the abdomen and are less inclined to be active, resulting in higher levels of oxidative stress and reduced oxidative capacity in the mitochondria.

One natural pathway to decrease the number of ROS produced in the cell is through uncoupled protons. The mitochondria inner membrane has numerous uncoupling membranes embedded in the surface that can act like doorways to be shut or partially open to decrease or increase the electrochemical gradient between the complexes (10). One way that you notice uncoupling working is it provides heat for your body. Here's a brief, entertaining video that explains uncoupling proteins (UPC) and how they work to keep us warm:

https://www.youtube.com/watch?v=2ommyLUshQg&t=38s
Credit: https://www.youtube.com/channel/UC67CzGY-98LwOAWca450UjQ

Uncoupling protein provides a way to control the use of respiratory protons for ATP generation as well as the metabolism of fatty acids (10). One negative implication of uncoupling proteins is that the controlled release of protons back into the mitochondrial matrix creates a decrease in ATP production but a continued consumption of oxygen and damage to membrane lipids (11). There's still ample room for research to understand the specific mechanisms of uncoupling proteins and their relationship to fatty-acid metabolism and ROS production, but uncoupling protons serve as a general link between metabolic pathways and insulin secretion relating to blood sugar control.

Insulin Resistance from Mitochondrial Dysfunction

Aging and dysfunction of the mitochondria have been closely associated with subsequent insulin resistance in many parts of the body, including the liver, pancreas, cardiovascular system, blood vessels, and in skeletal muscle (8). Cells that experience damage to their mitochondria and subsequently progress to diabetic symptoms are insulin-sensitive peripheral cells and islet beta cells in the pancreas (4). Several studies have shown that when mitochondrial function is compromised, insulin secretion shows a similar pattern of decline (4). Mitochondrial dysfunction has also been paired with an increase in the concentration of fatty-acid metabolites within the cell (3). When mitochondria age, their ability to facilitate oxidative phosphorylation slows, which, in turn, allows the buildup of intracellular fatty acids (3). These defects are characterized by fewer mitochondria within cells and increased genetic mutations that prevent the mitochondria from performing at full capacity (3).

When pancreatic beta cells are unable to sense increased glucose levels due to a diabetic condition, insulin isn't produced at adequate rates to normalize blood sugar levels (8). Defective mitochondria may contain abnormalities in their structure, genetic material, or subunits along the respiratory chain that decrease the output of beta oxidation (8). Additionally, mitochondrial dysfunction caused by excess lipids and ROS impairs insulin signaling pathways (8). The culmination of an influx in ROS brought about by genetic disposition, excessive caloric intake, physical inactivity, situations of hypoxia, and the natural aging process has dangerous implications for chronic diseases such as type 2 diabetes.

Lipid Peroxidation

Free-radical oxygenated species and excessive consumption of oxygen in the electron transport chain of the mitochondria lead to detrimental damage to surrounding genetic material, proteins, and lipids that are responsible for membrane structure and transport. As mentioned above, uncoupled protons can induce increased levels of oxygen consumption while redirecting protons away from ATP production. This mechanism and subsequent ROS production cause serious damage to lipids that make up the mitochondrial membrane (11). Obese and diabetic individuals have higher levels of fatty acids in their bodies, enhancing susceptibility to oxidative damage. Many lipids are located in close to mitochondria and are prime components of the mitochondrial membrane (4). Given this proximity, lipid peroxidation is of grave concern to diabetic patients with defective mitochondria.

Mitochondrial Glycation and Diabetes

There exist many abnormal mutations to metabolic pathway structures that impede the process of energy production, whether that be through genetic mutations that express the wrong structures for proteins or through oxidative stress, which can damage proteins and genetic material in the mitochondria. Another commonly proposed alteration to the mitochondria is glycation. Glycation is a process brought about by both oxidative and nonoxidative reactions to attach a sugar molecule to another structure such as a protein or a lipid (12). These processes are important to understand because attaching sugars to molecules in the metabolic pathway can cause serious malfunction and modification of structures such that their function is impaired and energy metabolism suffers. Various adverse effects of this alteration in structure include protein aggregation, alterations to protein conformation, and severe changes to membrane structure and function (12). The stability of the mitochondrial membrane is imperative for shuttling fatty acids into the mitochondria for beta oxidation and other metabolic substrates that help to sustain proper ATP levels in the body. If the lipids that make up the membrane are glycated, it's easier for lipid peroxidation to occur and fluidity of the membrane to increase (12). The glycation process supports pathways of oxidative stress and inflammation (12). Glycation processes are more frequently observed in diabetic patients compared with normal, healthy patients (12). More research is needed to develop strategies to prevent the addition of sugar derivatives to molecules involved in energy production, though some strategies have focused on the utilization of enzymes to remove byproducts of glycation.

Inability to Switch between Two Modes of Substrate Energy Utilization

There are two main substrates used for energy production: glucose and fatty acids. As detailed in previous chapters, fatty acids are a more efficient mode of energy production in the mitochondria because they produce more units of ATP per unit of substrate compared with glucose metabolism. It's essential that the body be able to switch between the two substrates for energy production because different situations require one or the other substrate. For example, during starvation, the muscles rely on fat storage for energy (4). When a person's caloric intake largely comprises carbohydrates, the metabolic pathway shifts to rely on glucose metabolism rather than beta oxidation. In cardiometabolic diseases such as type 2 diabetes and obesity, a "metabolic inflexibility" occurs when the body struggles to alternate between oxidative phosphorylation and glucose metabolism, resulting in increased insulin resistance (4). When the heart experiences high levels of stress in the diabetic condition, the body switches to glucose metabolism as a response (4).

There are certain benefits to each pathway of metabolism for the body, which are contingent on what the body needs most during that period. If there's more reliance on fats as the basis for energy production, the reliance on carbohydrates will subside and allow proper storage of carbohydrates within the body (11). If fats are underutilized as substrates, fat accumulates in the cell, which can exacerbate insulin resistance and lead to elevated levels of glycated lipids in the cell.

Therapeutic Interventions to Improve Blood Sugar Regulation

There are several interventions that target insulin resistance and improve control of blood sugar levels in diabetic and prediabetic patients. As we've shown, the number and well-being of mitochondria are vital to proper insulin function and signaling. Therefore, many of the available interventions to improve mitochondrial health focus on mitochondrial biogenesis and reduced oxidative stress.

To ameliorate mitochondrial dysfunction, the number and

function of existing mitochondria must be mediated by either fusing the functional portions of dysfunctional mitochondria, enhancing the biogenesis of new mitochondria, or removing malfunctioning components of mitochondria by the process of autophagy (11). The combination of diverse approaches—spanning dietary alterations, pharmacological agents, natural supplements, and lifestyle changes—will likely produce optimal results to treat this cardiometabolic disorder.

Pharmacological Agents

Thiazolidinediones (such as metformin) used in pharmacological studies of insulin resistance have shown improvements in the function of the liver, adipocytes, and the heart (8). In addition to the suppression of insulin resistance, these agents contribute to improvements in beta cell function in pancreatic cells responsible for insulin secretion and in endothelial function (8). Metformin is a medication used to improve insulin sensitivity and alleviate the influence of oxidative stress. Metformin reduces the production of ROS and allows a spike in mitochondrial biogenesis (8). An additional class of pharmacological agents used to improve blood sugar regulation comprises drugs associated with angiotensin. Angiotensin is a hormone that causes elevated blood pressure and vasoconstriction and supports pathways that lead to increased oxidative stress (13). Specific agents available to mitigate these harmful responses to the angiotensin II hormone include angiotensin-converting enzyme (ACE) inhibitors and angiotensin receptor blockers (8). These compounds can impede mechanisms that increase ROS production and dismantle mitochondrial energy production.

Natural Supplements

Antioxidants are an obvious intervention choice for insulin

resistance and defective mitochondria because of the large role that ROS play in blood sugar maintenance. In a reciprocal effect, defective mitochondria are found in cells with higher ROS production because they damage the genetic and structural components, and these malfunctioning mitochondria naturally produce higher levels of ROS (8). Antioxidants are able to impede these harmful oxygenated byproducts and preserve mitochondria.

Alpha Lipoic Acid

Alpha-lipoic acid (ALA) is a prominent antioxidant and anti-inflammatory agent that's been shown to remedy numerous chronic diseases—in particular, diabetes. ALA has been shown to lower glucose concentrations in the blood and quench ROS that inhibit proper insulin signals (8). The role of ALA in glucose metabolism and its antioxidant capabilities to remove harmful ROS from the cell support its use in the management of blood sugar levels among diabetic patients.

Coenzyme Q10

Coenzyme Q10 (CoQ10) is an integral component in the mitochondria along the electron transport chain, serving as a key antioxidant in the reduction-oxidation reactions that occur to support ATP production. As a strong antioxidant, CoQ10 plays an important role in pathways related to metabolism, signaling, and transport (11). The mitochondrial electron transport chain must pass electrons among the four complexes and ultimately hand them off to oxygen. CoQ10, as an antioxidant, is able to undergo reduction to gain an electron as well as oxidation to give up an electron (11). CoQ10 is a beneficial supplement in chronic conditions, such as diabetes, because it supports energy generation and proper functionality of the electron transport chain to return to a state of proper mitochondrial function. In clinical studies with CoQ10 supplementation, CoQ10 was able to lower fasting and postprandial glucose levels to improve insulin secretion and blood sugar control (4). In studies of mitochondrial defects passed from mother to child, CoQ10 is considered a potential remedy, demonstrating promising results in early studies (4).

L-carnitine

Found naturally in the body, L-carnitine is an essential compound whose most prominent role involves fatty-acid transportation. L-carnitine serves the metabolic process by shuttling fatty acids into the mitochondria to undergo beta oxidation and maintaining proper levels of coenzyme A (11). Supplementation of L-carnitine can provide tremendous benefits to diabetic patients, especially if it's deficient in the body. When levels of L-carnitine are low, the diabetic condition is exacerbated from lessened mitochondrial output and worsening insulin resistance (11). In diabetes, fatty-acid oxidation is impaired. L-carnitine supplementation supports the regenerating of these metabolic pathways.

NADH

Nicotinamide adenine dinucleotide (NADH) is a primary metabolite used in mitochondrial energy production, and it is a source of natural supplementation to remedy mitochondrial dysfunction. Once reduced, NADH loses electrons to the electron transport chain, which ultimately end up with molecular oxygen and support the proton pump that produces ATP from ADP (11). NADH is also an antioxidant when it gains an electron (reduction), assisting to quench free-radical species to prevent further oxidative damage to mitochondrial proteins, membranes, and genetic material (11). Additional research is needed to confirm its success as an oral supplement but, based on cellular mechanisms, NADH could be a key player in the health of mitochondria and have an influence on chronic disease.

Physical Activity

A widely-accepted solution to insulin resistance, along with many of the adverse symptoms of chronic diseases, is a consistent exercise regimen. The recommended treatment for type 2 diabetes is partnering moderate-intensity physical activity with a 5 to 10 percent weight loss (14). Human and animal models have exhibited beneficial effects on glucose and insulin levels in subjects that undergo regimented exercise programs (8). Studies have shown that patients who lose weight observe significant benefits in their cardiometabolic health. One such study found that, among patients who lost weight, there was an increase in mitochondrial density and insulin-sensitivity improvements (15).

Patients in this study accomplished an average weight loss of 7.1 percent and a decrease in mean HbA1c levels from 7.9 to 6.5 percent (15). In addition, diabetic patients who lost weight also observed decreases in fasting and postprandial blood glucose levels (15). Studies like this demonstrate the notable improvements in glucose levels and other diabetic complications resulting from weight loss.

Physical inactivity remains an integral component in the development of diabetes. Prolonged physical activity has the potential to mediate insulin resistance and glycogen synthesis in muscle cells, further improving glucose transport and phosphorylation (3). In addition to normalized insulin sensitivity and glucose processing, physical activity promotes healthy weight loss, which improves the prognosis of type 2 diabetes and other cardiometabolic disorders.

The process of gluconeogenesis is likely the force that drives increased glucose production in diabetic patients. Physical activity is proposed to slow down the rates of gluconeogenesis and, in turn, return cellular glucose levels to equilibrium (3). In terms of energy production, exercise stimulates energy levels by supporting the generation of new mitochondria that return the cell to normal (8).

Not only does the number of mitochondria improve, but the metabolic processing of glucose and the size of the mitochondria also return to equilibrium (8). One study on exercise and diabetic symptoms found that patients on a moderate-intensity exercise regimen maintained an average increase in mitochondrial density of 67 percent, with coinciding increases in mitochondrial size (14).

When energy consumed outweighs energy expended, individuals face inevitable weight gain and eventually become obese. Obesity contributes to insulin resistance through fat accumulation. Excess body fat has been shown to elicit a variety of negative health effects, with fat in the abdominal region driving insulin resistance more than fat stored in a pear shape around the hips (3).

Nutritional Strategies

Obesity and diabetic blood sugar irregularities are undeniably intertwined. Thus, caloric restriction implemented in the treatment of type 2 diabetes improves glucose metabolism and insulin sensitivity. By the means of Nitric Oxide production, restricting the input of calories promotes the generation of mitochondria and the oxidative capabilities of these structures (7).

Another disease category where mitochondrial dysfunction is

implicated is neurological disorders, which are discussed in the next chapter.

References:

1. Srivastava, A.K. "Challenges in the Treatment of Cardiometabolic Syndrome." *Indian Journal of Pharmacology* 44, no. 2 (2012): 155–6. doi:10.4103/0253-7613.93579.
2. Patti, M.E. et al. "The Role of Mitochondria in the Pathogenesis of Type 2 Diabetes." *Endocrine Reviews* 31 (2010): 364–95.
3. Parish, R. and Petersen, K.F. "Mitochondrial Dysfunction and Type 2 Diabetes." *Current Diabetes Reports* 5, no. 3 (2005): 177–83.
4. Sivitz, W.I. and Yorek, M.A. "Mitochondrial Dysfunction in Diabetes: From Molecular Mechanisms to Functional Significance and Therapeutic Opportunities." *Antioxidants and Redox Signaling* 12.4 (2010): 537–77.
5. Coelho, G.D.P., Martins, V.S., do Amaral, L.V., Novaes, R.D., Sarandy, M.M., and Gonçalves, R.V. "Applicability of Isolates and Fractions of Plant Extracts in Murine Models in Type II Diabetes: A Systematic Review." *Evidence-based Complementary and Alternative Medicine* (*eCAM*) 2016, no. 3537163 (2016). doi:10.1155/2016/3537163.
6. Shulman, G.I., Rothman, D.L., Jue, T., et al. "Quantitation of Muscle Glycogen Synthesis in Normal Subjects and Subjects with Non-Insulin-Dependent Diabetes by 13C Nuclear Magnetic Resonance Spectroscopy." *New England Journal of Medicine* 322 (1990): 223–8.
7. Reza, H., Bonavaud, S.M., Armstrong, J.L., McCormack, J.G., and Yeaman, S.J. "Control of Glycogen Synthesis by Glucose, Glycogen, and Insulin in Cultured Human Muscle Cells." *Diabetes* 50, no. 4 (2001): 720–6. doi: 10.2337/diabetes.50.4.720.
8. Kim, J., Wei, Y., and Sowers, J.R. "Role of

Mitochondrial Dysfunction in Insulin Resistance." *Circulation Research* 102, no. 4 (2008): 401–14. doi:10.1161/CIRCRESAHA.107.165472.

9. Sparks, L.M., Xie, H., Koza, R.A., Mynatt, R., Hulver, M.W., Bray, G.A., and Smith, S.R. "A High-Fat Diet Coordinately Downregulates Genes Required for Mitochondrial Oxidative Phosphorylation in Skeletal Muscle." *Diabetes* 54, no. 7 (July 2005): 1926–33. doi: 10.2337/diabetes.54.7.1926.

10. Rousset, S., Alves-Guerra, M.C., Mozo, J., Miroux, B., Cassard-Doulcier, A.M., Bouillaud, F., and Ricquier, D. "The Biology of Mitochondrial Uncoupling Proteins." *Diabetes* 53, supplement 1 (February 2004): S130–5. doi:10.2337/diabetes.53.2007.S130.

11. Nicolson G.L. "Mitochondrial Dysfunction and Chronic Disease: Treatment with Natural Supplements." *Integrative Medicine: A Clinician's Journal* 13, no. 4 (2014): 35–43.

12. Pun, P.B.L. and Murphy, M.P. "Pathological Significance of Mitochondrial Glycation." *International Journal of Cell Biology* 2012, article ID 843505 (2012). doi:10.1155/2012/843505.

13. Dikalov, S.I. and Nazarewicz, R.R. "Angiotensin II-Induced Production of Mitochondrial Reactive Oxygen Species: Potential Mechanisms and Relevance for Cardiovascular Disease." *Antioxidants and Redox Signaling* 19, no. 10 (2013): 1085–94. doi:10.1089/ars.2012.4604.

14. Toledo F.G. et al. "Effects of Physical Activity and Weight Loss on Skeletal Muscle Mitochondria and Relationship with Glucose Control in Type 2 Diabetes." *Diabetes* 56 (2007): 2142–7.

15. Muoio, D.M. and Newgard, C.B. "Mechanisms of Disease: Molecular and Metabolic Mechanisms of Insulin Resistance and Beta-Cell Failure in Type 2 Diabetes." *Nature Reviews Molecular Cell Biology* 9,

no. 3 (March 2008): 193–205.

Chapter 7

Mitochondrial Dysfunction and Neurodegenerative Diseases

As we've discussed in previous chapters, mitochondrial dysfunctions can be an early stage of several mitochondrial disorders, including neurodegenerative diseases, cancers, and cardiovascular diseases. Mitochondrial dysfunctions may be caused by mutations to mitochondrial DNA or mutations to nuclear DNA that encode for mitochondrial component production (1). In Chapter 4, "Mitochondria and How They Make Energy," we discussed how, during evolution, the great majority of mitochondrial genes were relocated into the safe harbor of the nucleus, but thirteen genes had to remain inside the mitochondria to participate in oxidative phosphorylation (OXIPHOS). These thirteen genes provide instructions for making enzymes involved in the OXIPHOS electron transport chain that creates ATP, the universal energy molecule of the cell. Any mutations to these genes make them inactive or less efficient in producing ATP, an outcome that results in mitochondrial dysfunction.

There's growing evidence that mitochondrial dysfunctions are responsible for many neurodegenerative diseases, including Alzheimer's disease, Parkinson's disease, and Huntington's disease. Defects in mitochondrial metabolism and, particularly, in the electron transport chain, may play a major role in the pathogenesis of these diseases. In addition to not producing sufficient ATP, structurally and functionally damaged mitochondria are predominant in producing pro-apoptotic factors and reactive oxygen species that result from an inefficient handling of electron transport. This results in increased leakage of high-energy electrons (2).

Excitotoxicity and apoptosis are major causes of neuronal cell damage and cell death. Excitotoxicity is the overstimulation of brain cell receptors by glutamate, a neurotransmitter. Excessive glutamate stimulation can cause apoptosis of brain cells. Dysfunctional mitochondria are both necessary and sufficient for these events to take place (1, 7).

The first evidence of mitochondrial involvement in the pathogenesis of neurodegenerative diseases was reported when Complex I deficiency was detected in the platelet mitochondria and substantia nigra of Parkinson's disease patients. Complex I is the critical first step in OXIPHOS, in which the enzyme ubiquinone oxidase synthesizes NADH from the Krebs cycle and transports protons across the inner mitochondrial membrane to support ATP synthesis (4, 5). Deficiencies in the mitochondria of Alzheimer's patients were first found in Complex I and Complex IV. Complex IV, cytochrome oxidase, is the last enzyme in the respiratory electron transport chain of the mitochondria described in Chapter 4. In Huntington's disease, evidence of mitochondrial deficiencies was discovered in Complex II, Complex III, or both. Complex II deficiency is caused by mutations to the mitochondrial genes encoding for succinate dehydrogenase.

Complex III deficiency is caused by mitochondrial mutations of genes that encode for cytochrome b, which oxidizes ubiquinol and reduces cytochrome c. Complex III is already a major source of electron leakage, producing superoxide radicals even during normal functioning. When mutations to mitochondrial Complex III genes occur, even more leakage of high-energy electrons takes place, which results in greater free-radical damage (6).

The energy demands of the brain are high—the brain is

responsible for 20 percent of the body's total oxygen consumption. The brain's energy needs are mostly driven by the neurons' requirement for energy. This intense energy need is continuous and, if the requirement isn't met, there are deadly consequences, as can be seen following even brief periods of oxygen or glucose deprivation (7). Neurons themselves have limited glycolytic capacity, which makes them highly dependent on OXIPHOS for their energetic needs (7). However, as we discussed in the Chapter 4, OXIPHOS is a major source of free radicals in the form of high-energy electrons that are byproducts of normal cellular respiration. The electron transport chain is somewhat inefficient, releasing high-energy electrons from Complex I, Complex III, and PMET (plasma membrane electron transport), where they react with molecular oxygen to yield superoxide radicals.

The Brain's Unique Vulnerability to Free-Radical Damage—The Perfect Storm

The brain is particularly vulnerable to free-radical-induced damage because of its high bioelectric activity, low antioxidant capacity, and high fat content (1). Neurons are post-mitotic cells formed during embryogenesis as well as in early growth and developmental stages. Mitotic cell populations are continually dividing and don't accumulate oxidative debris such as lipofuchsin, beta amyloid, or tau proteins, which are all associated with neurodegenerative diseases. However, as post-mitotic cells, neurons accumulate lipofuchsin, beta amyloid, and tau proteins in abundance (8).

The high fat content of the brain (the brain is essential made from cholesterol and has its own private reserve of cholesterol) is an ideal playing field for the propagation of free radicals, where a high-energy electron from mitochondrial transport leakage or PMET removes a hydrogen from a fatty acid. The fatty acid is then converted into a peroxyl radical by reacting with the always-present oxygen in cells. This peroxyl radical then steals another hydrogen from an adjacent molecule to pair its unpaired electron. This process of abstracting a hydrogen continues until eight to ten molecules of the fatty acid are destroyed. The termination of this free-radical cascade occurs only when an antioxidant donates a hydrogen atom or an electron. This entire process is called lipid peroxidation and results in cell and organelle membrane damage in the brain (9).

Amyloid Aggregation

Beta amyloid is a high-molecular-weight, oxidized protein mass that's a direct product of disintegrated pieces of cell and organelle membranes and misfolded proteins. Leakage of high-energy electrons from defective mitochondria causes protein misfolding (1, 3, 7, 9, 10, 11).

The term "misfolded proteins" refers to proteins that are partially or completely devoid of the conformation associated with optimal stability and biological function and that are essentially inactive in the cell. These misfolded proteins aggregate into assemblies and/or interact inappropriately with other cellular components, impairing cell viability and function and eventually causing cell death (10, 11).

Amyloid aggregation is a type of protein misfolding occurring in the brain and, as discussed in Chapter 2 on molecular machines, space is at a premium in all cells because of the increasingly crowded cellular conditions resulting from oxidized protein buildup. Amyloid aggregation is a hallmark of neurodegenerative diseases affecting the brain. Aggregated amyloid forms spread in the brains of patients suffering from neurodegenerative diseases such as Alzheimer's disease, Parkinson's disease, and Huntington's disease (1, 2, 3, 7, 9).

Studies of the brain cells of Alzheimer's patients demonstrate that free-radical-induced lipid peroxidation is widespread and leads to amyloid aggregation (10, 11, 12).

Once formed, amyloid aggregation causes increased lipid peroxidation in brain cell membranes; its formation is inhibited by an array of mitochondrially targeted antioxidants (12, 17, 18). Amyloid aggregation leads to further irritation of nearby cells and exacerbates intracellular damage in an endless cycle, finally resulting in cell death (10, 11, 12).

LIFESTYLE FACTORS AND NEURODEGENERATION

The Discovery of the Brain's Glymphatic System

The brain, as part of the central nervous system, contains blood vessels, but it was long believed to lack a lymphatic vessel drainage system. However, researchers recently discovered a series of channels that surround blood vessels within the brains of mice. This system, managed by the brain's glial cells, has been termed the glymphatic system. The glymphatic system moves cerebrospinal fluid—a clear liquid surrounding the brain and spinal cord—quickly and deeply throughout the brain, removing waste during sleep. This waste includes excreted beta amyloid proteins (19, 21, 22).

The discovery of a pathway for immune cells to exit the central nervous system raises the question of whether disruption of this route may be involved in neurological disorders associated with immune system dysfunction, such as multiple sclerosis, meningitis, and Alzheimer's disease (19, 20, 21, 22). To test the effects that sleep has on the glymphatic clearance of cellular waste, scientists observed the difference between the glymphatic system's activity during waking hours and during sleep (20, 21).

Using a mouse model, researchers demonstrated that sleep helps restore the brain by flushing out toxins that build up during waking hours. They reported that the glymphatic system can help remove beta amyloid from brain tissue, suggesting a potential new role for sleep in health and disease.

The importance of sleep cannot be overemphasized, especially its potential to prevent cognitive decline or neurodegenerative disorders.

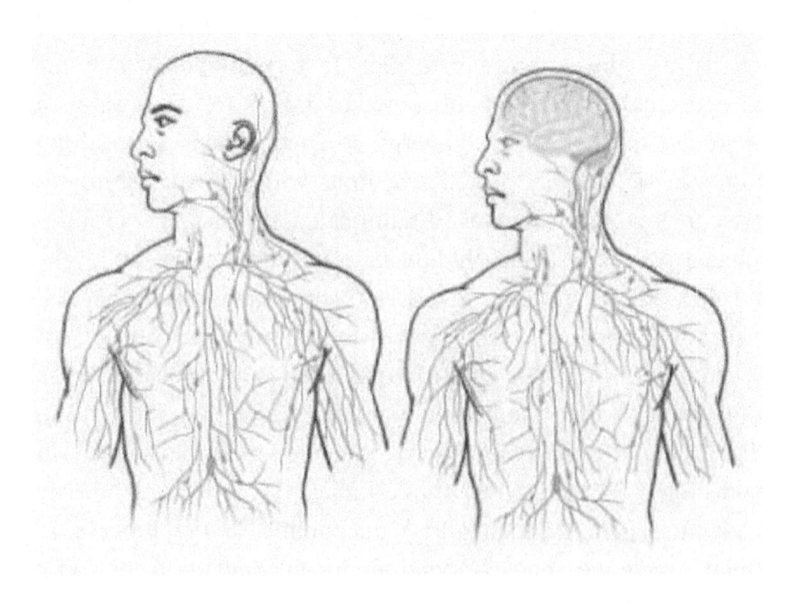

Figure 5. Maps of the body's lymphatic system: old (left) and updated (right) to reflect the new discovery of its presence in the brain.
Courtesy: University of Virginia Health System.

Calorie Restriction and Neurodegeneration

Calorie restriction is the only universally consistent method for increasing animal lifespan across species. The effects of this regimen have been studied since 1930 (23). Restricting carbohydrates in humans lowers body weight more than a low-fat diet of the same caloric content. There's a greater decrease in serum triglyceride levels and a greater increase high-density lipoprotein (HDL, the "good" cholesterol) levels in a low-carbohydrate diet than in a low-fat diet (24).

Ketone bodies, including β-hydroxybutyrate, are produced by the consumption of low-carbohydrate, ketogenic diets. Ketone bodies serve as an alternate energy source in states of metabolic stress—namely, dieting—and contribute to the neuroprotective activity of a low-carbohydrate diet. In fact, β-hydroxybutyrate may provide a more efficient source of energy for the brain than glucose (25). The enhanced energy production capacity resulting from a ketogenic diet provides neurons with a greater ability to resist metabolic challenges. Additionally, biochemical changes induced by this diet—including ketosis, high serum-fat levels, and low-serum glucose levels—protect against neuronal-cell death by apoptosis and necrosis through a variety of mechanisms, including mitochondrial uncoupling as well as antioxidant and anti-inflammatory effects (26, 27). Mitochondrial uncoupling improves mitochondrial efficiency in producing ATP and lowers free-radical levels during energy production (27). Mitochondrial uncoupling is the process in which oxidative phosphorylation is uncoupled from ATP synthesis, with the resulting energy dissipated as heat (26).

Exercise and Neurodegeneration

In both animal and human studies lasting up to nine months, long-term, regular aerobic exercise was shown to improve cognition, lower dementia risk, and slow the progression of dementia. The potential benefits accrue with long-term, regular exercise sufficient to increase the heart rate and the need for oxygen, with demonstrable improvements in mitochondrial function (28, 29). Twenty-nine human trials have measured the effects of exercise in adults without dementia, and three trials have measured exercise effects on patients with dementia. Results included improved scores in memory, attention, processing speed, and executive function. While brain gray-matter volumes decrease with advancing age, as seen in MRI brain images, aerobic exercise increases brain gray matter (28, 30). Aerobic exercise has been shown to reduce brain pathogenic beta-amyloid plaque in most but not all studies (31, 32).

Hypoxia and Neurodegeneration

Various cardiorespiratory disorders can cause periods of chronic hypoxia that place people at greater risk of developing dementia, especially Alzheimer's disease. Cardiac arrest is another source of hypoxia in humans and, over time, causes an increase in detectable levels of beta amyloid in the blood (33, 34, 35). The deposit of beta amyloid in the brain has also been shown to occur after severe head injury.

Chronic stress leads to shallow breathing and arterial stiffness, resulting in an impaired flow of nutrients and O_2 to the brain. One study showed that older women with breathing disorders that occur during sleep have a greater risk of developing dementia and cognitive impairment (36).

Therapeutic Interventions to Improve Neurological Health

As we've shown, the number and well-being of mitochondria are vital to proper brain function and signaling. Therefore, many of the available interventions to improve mitochondrial health focus on mitochondrial biogenesis and reduced oxidative stress. To ameliorate mitochondrial dysfunction, the number and function of existing mitochondria must be mediated by either fusing the functional portions of dysfunctional mitochondria, enhancing the biogenesis of new mitochondria, or removing malfunctioning components of mitochondria through autophagy. The combination of diverse approaches—spanning dietary alterations, natural supplements, and lifestyle changes—will likely produce optimal results for improving neurological outcomes.

D-ribose

D-ribose is a unique mitochondrial energy source because it's the sugar moiety (part of a molecule) of ATP and can be incorporated directly into mitochondria both for direct energy and ATP production—without having to be converted into intermediate compounds (37). D-ribose is a conditionally essential nutrient that acts as a metabolic supplement for the heart under certain pathologic cardiac conditions in which the nucleotides ATP, ADP, and AMP have been degraded and lost. The heart's ability to resynthesize ATP is limited by the supply of D-ribose, which is a necessary component of the adenine nucleotide structure (37).

Acetyl-L-Carnitine

Acetyl-L-carnitine is a natural compound found in the human body that restores function to aging mitochondria. Acetyl-carnitine transferase levels decline in older cells. This is the key enzyme that transports carnitines across mitochondrial membranes. Dietary acetyl-L-carnitine restores mitochondrial function to more youthful energy levels in older animals with age-related mitochondrial dysfunction. Acetyl-L-carnitine is synergistic with alpha lipoic acid in restoring mitochondrial function (38, 39).

In humans, acetyl-L-carnitine can pass through the blood-brain barrier and restore neuronal function. In meta-analyses of studies, acetyl-L-carnitine has been shown to relieve neuropathic pain, treat depression, evidence antifatigue properties, and enhance cognitive function. Acetyl-L-carnitine slows the progression of Alzheimer's disease. Supplemental doses usually range from five hundred to two thousand milligrams (mg) per day (40, 41, 42, 43).

Astragaloside IV

Astragaloside IV (AG-IV) is the most abundant saponin (glucoside) found in Chinese skullcap, Astragalus membranaceus. Astragaloside IV enhances cell viability and decreases the accumulation of mitochondrial superoxide and intracellular reactive oxygen species (ROS) in cells that already contain large amounts of beta amyloid. Beta amyloid generates free radicals and causes mitochondrial dysfunction (44).

In cells treated with beta amyloid, AG-IV improves mitochondrial function, maintains mitochondrial membrane potential, and suppresses the release of caspase-3, the executioner gene that lowers superoxide dismutase production in mitochondria, causing apoptosis (44). Beta amyloid causes mitochondrial dysfunction by a deficiency of glucose metabolism and deactivation of key enzymes that are vital to oxidative phosphorylation (45, 46).

Curcumin

Curcumin is a mixture of curcuminoids found in the same ratio that occurs naturally in turmeric rhizomes (aerial roots). Curcuminoids cross the blood-brain barrier. In Alzheimer's disease mouse models, curcumin reduces beta-amyloid plaque burden; in in vitro studies, it reduces beta amyloid-positive proteins (47, 48).

Mitochondrial function and cell viability are increased in curcumin-treated cells. In studies, curcumin pre- and post-treated cells incubated with beta amyloid showed less mitochondrial dysfunction and maintained cell viability and mitochondrial dynamics, including mitochondrial biogenesis and synaptic activity. The protective effects of curcumin were stronger in pretreated cells than in post-treated cells, indicating that curcumin works better at prevention than in the treatment of existing Alzheimer's disease-like neurons (49).

A 2008 clinical trial demonstrated that curcumin has beneficial effects on Alzheimer's disease patients. Despite the promising prospects, the exact mechanism(s) by which curcumin exerts its neuroprotection remains unknown (50).

Curcumin significantly improved memory deficits in a mouse model of Alzheimer's disease and promoted neuronal function in vivo and in vitro. Curcumin also reduced inflammatory cytokine production and inhibited nuclear factor kappa B (NF-κB), the beginning of the inflammatory pathway, suggesting the beneficial effects of curcumin on Alzheimer's disease are due to the suppression of neuroinflammation (5).

EGCG

Epigallocatechin gallate (EGCG) is a major flavonoid in green tea and has been shown to reduce both beta amyloid production in cultured cells and amyloid plaque buildup in a transgenic mouse model of Alzheimer's (52, 53). A study aimed at investigating the association between green tea consumption and cognitive function in elderly Japanese subjects showed that a higher consumption of this beverage was associated with a lower prevalence of cognitive impairment (54).

Octyl Gallate

The authors of an EGCG study on Alzheimer's disease in a mouse model screened natural EGCG-like compounds and found that low doses of octyl gallate drastically decreased beta amyloid generation by increasing the disposal, or proteolysis, of the amyloid precursor molecule APP (amyloid precursor protein) in neuron-like cells. When tested in an Alzheimer's disease mouse model over one week, a daily dose of octyl gallate decreased beta-amyloid levels that are associated with lower APP—an outcome achieved by activating the enzyme APP-a secretase, which dissolves APP (55).

Resveratrol

Resveratrol has received considerable attention in the context of neurodegenerative diseases because of its antineuroinflammatory activity and its ability to inhibit amyloid formation, the precursor peptide linked to amyloid plaques. Activation of the microglia—the macrophages that reside in the brain near amyloid plaques—is a key hallmark of Alzheimer's disease. Recent evidence in mouse models indicates that microglia are required for the neurodegenerative process of Alzheimer's (56). Orally administered resveratrol in a mouse model of cerebral amyloid deposition lowered microglial activation associated with amyloid plaque formation (57).

Certain fragments of the beta amyloid protein, when broken down by the protease digestive enzymes, retain the pathophysiology of beta amyloid. These fragments have been shown to cause mitochondrial dysfunction. Resveratrol was shown to dose-dependently inhibit beta-amyloid breakdown products by depolymerizing them, leading to the formation of nontoxic proteins (58).

Myricetin and Quercetin

Inhibition of amyloid aggregation has been highlighted as one of the potential targets for the development of new drugs to treat

Alzheimer's disease. Several reports have discussed the role of myricetin (a flavonoid found in fruits, vegetables, tea, and red wine) in the inhibition of beta amyloid assembly (59). The protective effects of myricetin also stem from the effect of the compound against tau proteins (59). A third mode of action that has been investigated widely is the ability of myricetin to block Alzheimer's-associated beta amyloid fibril formation (60, 61).

Quercetin is the major neuroprotective compound in coffee. Coffee consumption has been shown to lower the risk of Parkinson's disease and Alzheimer's disease in humans, but only recently have the active components been identified (62). In one study, quercetin reduced oxidative and nitrative (peroxynitrite radical) damage to DNA as well as to the cell membranes and proteins in SH-SY5Y brain cells. Quercetin caused a significant increase in glutathione, which becomes depleted in SH-SY5Y cells exposed to neurotoxins released by the astrocyte and microglia immune cells in the brain. The authors of the study concluded, "The data indicate that quercetin is the major neuroprotective component in coffee against Parkinson's disease and Alzheimer's disease" (63).

Quercetin has been shown in several studies to be a potent inhibitor of amyloid aggregation and amyloid toxicity. Quercetin is very similar in structure to myricetin (64).

Gouteng (Uncaria rhynchophylla)

Uncaria species (*gou teng* in Chinese) have been used as ethnopharmacological medicines (indigenous or folk medicines) to treat cardiovascular and cerebral diseases (65). Gou teng enhances the autophagy pathway, a type of programmed cell death that clears dysfunctional cells that contain aggregated amyloid proteins. Mice without the genes needed for autophagy quickly develop Parkinson's-like symptoms (66).

Glutamate is a neurotransmitter in normal neuronal cells. However, high concentrations of glutamate produce oxidative stress, resulting in the neurodegeneration seen in Alzheimer's disease and Parkinson's disease. High glutamate levels are also produced in stroke, where blood flow is interrupted. Alkaloids found in gou teng protect brain cells against glutamate-induced cell death (67).

Creatine

Creatine is a naturally occurring compound found in muscles and is assembled from the amino acids methionine, glycine, and arginine (68). Creatine phosphate can donate a phosphate group to ADP to form ATP and can be used to regenerate ATP in storage. Creatine can protect brain cells from neurotoxins and act as a therapeutic after ischemic events such as stroke (69). Creatine can be used to treat muscular disorders, such as those seen in ALS, Parkinson's disease, Huntington's disease, and muscular dystrophy (70, 71, 72).

Creatine has also been used to treat heart failure. It's found in excitable cells such as cardiomyocytes, where it plays an important role in the transport of mitochondrial energy to ensure that supply meets the ever-changing demands of the heart. Multiple components of cardiac mitochondria, including intracellular creatine levels, are reduced in heart failure (73).

In the next chapter, we'll discuss the influence of mitochondrial dysfunction on arthritis.

References:

1. Hroudová, J., Singh, N., and Fišar Z. "Mitochondrial Dysfunctions in Neurodegenerative Diseases: Relevance to Alzheimer Disease." *BioMed Research International* 2014, article ID 175062 (2014).

2. Moreira, P.I., Zhu, X., Wang, X., Lee, H., Nunomura, A., Petersen, R.B., Perry, G., and Smith, M.A. "Mitochondria: A Therapeutic Target in Neurodegeneration." *Biochimica et Biophysica Acta (BBA) – Molecular Basis of Disease*. 1802 (2010): 212–20.

3. Emerit, J., Edeas, M., and Bricaire, F. "Neurodegenerative Diseases and Oxidative Stress." *Biomedicine and Pharmacotherapy* 58, no. 1 (2004): 39–46.

4. Schapira, A.H.V., Cooper, J.M., Dexter, D., Clark, J.B., Jenner, P., and Marsden, C.D. "Mitochondrial Complex I Deficiency in Parkinson's Disease." *Journal of Neurochemistry* 54, no. 3 (1990): 823–7.

5. Parker Jr., W.D., Boyson, S.J., and Parks, J.K. "Abnormalities of the Electron Transport Chain in Idiopathic Parkinson's Disease." *Annals of Neurology* 26, no. 6 (1989): 719–23.

6. Moran, M., Moreno-Lastres, D., Marin-Buera, L., Arenas, J., Martin, M.A., and Ugalde, C. "Mitochondrial Respiratory Chain Dysfunction: Implications in Neurodegeneration." *Free Radical Biology and Medicine* 53 (2012): 595–609.

7. Moreira, P., Zhu, X., Wang, X., et al. "Mitochondria: A Therapeutic Target in Neurodegeneration." *Biochimica et Biophysica Acta – Molecular Basis of Disease* 1802, no. 2 (2010): 212–20.

8. Okabe, S., Nilssen, K.F., Spiro, C., et al. "Development of Neuronal Precursor Cells and Functional Post-Mitotic Neurons from Embryonic Stem Cells in Vitro."

Mechanisms of Development 59 (1996): 89–102.

9. Butterfield, D. and Lauderback, C. "Lipid Peroxidation and Protein Oxidation in Alzheimer's Disease." *Free Radical Biology and Medicine* 32, no. 11 (2002): 1050–60.

10. Stefani, M. and Dobson, C.M. "Protein Aggregation and Aggregate Toxicity: New Insights into Protein Folding, Misfolding Diseases, and Biological Evolution." *Journal of Molecular Medicine* 81, no. 11 (November 2003): 678–99.

11. Chiti, F. and Dobson, C.M. "Protein Misfolding, Functional Amyloid, and Human Disease." *Annual Review of Biochemistry* 75 (2006): 333–66.

12. Butterfield, D.A., Hensley, K., Harris, M., Mattson, M., and Carney, J. "Beta-Amyloid Peptide Free Radical Fragments Initiate Synaptosomal Lipoperoxidation in a Sequence-Specific Fashion: Implications to Alzheimer's Disease." *Biochemical and Biophysical Research Communications* 200 (1994): 710–5.

13. Daniels, W.M., van Rensburg, S.J., van Zyl, J.M., and Taljaard, J.J. "Melatonin Prevents Beta-Amyloid Induced Lipid Peroxidation." *Journal of Pineal Research* 24 (1998): 78–82.

14. Gridley, K.E., Green, P.S., and Simpkins, J.W. "Low Concentrations of Estradiol Reduce Beta-Amyloid (25–35)-induced Toxicity, Lipid Peroxidation, and Glucose Utilization in Human SK-N-SH Neuroblastoma Cells." *Brain Research* 778 (1997): 158–65.

15. Bruce-Keller, A.J., Begley, J.G., Fu, W., Butterfield, D.A., Bredesen, D.E., Hutchins, J.B., Hensley, K., and Mattson, M.P. "Bcl-2 Protects Isolated Plasma and Mitochondrial Membranes Against Lipid Peroxidation Induced by Hydrogen Peroxide and Amyloid Beta-Peptide." *Journal of Neurochemistry* 70 (1998): 31–9.

16. Mark, R.J., Fuson, K.S., and May, P.C. "Characterization of 8-epiprostaglandin F2 as a Marker

of Amyloid-Peptide-Induced Oxidative Damage." *Journal of Neurochemistry* 72 (1999): 1146–53.

17. Mark, R.J., Lovell, M.A., Markesbery, W.R., Uchida, K., and Mattson, M.P. "A Role for 4 Hydroxynonenal, an Aldehydic Product of Lipid Peroxidation, in Disruption of Ion Homeostasis and Neuronal Death Induced by Amyloid-Peptide." *Journal of Neurochemistry* 68 (1997): 255–64.

18. Avdulov, N.A., Chochina, S.V., Igbavboa, U., O'Hare, E.O., Schroeder, F., Cleary, J.P., and Wood, W.G. "Amyloid Beta-Peptides Increase Annular and Bulk Fluidity and Induce Lipid Peroxidation in Brain Synaptic Plasma Membranes." *Journal of Neurochemistry* 68 (1997): 2086–91.

19. Torgan, C. "Lymphatic Vessels Discovered in Central Nervous System." *NIH Research* Matters. June 15, 2015. https://www.nih.gov/news-events/nih-research-matters/lymphatic-vessels-discovered-central-nervous-system.

20. "How Sleep Clears the Brain." *NIH Research Matters*. October 28, 2013.

21. Louveau, A., Smirnov, I., Keyes, T.J., Eccles, J.D., et al. "Structural and Functional Features of Central Nervous System Lymphatic Vessels." *Nature* 523, no. 7560 (July 16, 2015): 337–41.

22. Iliff, J., Goldman, S., and Nedergaard, M. "Clearing the Mind: Implications of Dural Lymphatic Vessels for Brain Function." *The Lancet Neurology* 14, no. 10 (October 2015): 977–9.

23. Al-Regaiey, K.A. "The Effects of Calorie Restriction on Aging: A Brief Review." *European Review for Medical and Pharmacological Sciences* 20, no. 11 (June 2016): 2468–73.

24. Yancy, W., Olsen, M.K., and Guyton, J.R. "A Low-Carbohydrate, Ketogenic Diet versus a Low-Fat Diet to Treat Obesity and Hyperlipidemia." *Annals of Internal*

Medicine 140 (2004): 769–77.

25. Gasior, M., Rogawski, M.A., and Hartman, A.L. "Neuroprotective and Disease-Modifying Effects of the Ketogenic Diet." *Behavioural Pharmacology* 17, no. 5-6 (September 2006): 431–9. Review.
26. Sullivan, P.G., Rippy, N.A., Dorenbos, K., Concepcion, R.C., Agarwal, A.K., and Rho, J.M. "The Ketogenic Diet Increases Mitochondrial Uncoupling Protein Levels and Activity." *Annals of Neurology* 55 (2004): 576–80.
27. Veech, R.L. "The Therapeutic Implications of Ketone Bodies: The Effects of Ketone Bodies in Pathological Conditions: Ketosis, Ketogenic Diet, Redox States, Insulin Resistance, and Mitochondrial Metabolism." *Prostaglandins, Leukotrienes, and Essential Fatty Acids* 70 (2004): 309–19.
28. Ahlskog, J., Geda, Y., Graff-Radford, N., and Petersen, R.C.. "Physical Exercise as a Preventive or Disease-Modifying Treatment of Dementia and Brain Aging." *Mayo Clinic Proceedings* 86, no. 9 (September 2011): 876–84.
29. Barbieri, E., Agostini, D., Polidori, E., et al. "The Pleiotropic Effect of Physical Exercise on Mitochondrial Dynamics in Aging Skeletal Muscle." *Oxidative Medicine and Cellular Longevity* 917085 (2015).
30. Erickson, K., Leckie, R., and Weinstein, A. Physical Activity, Fitness, and Gray Matter Volume." *Neurobiology of Aging* 35, supplement 2 (September 2014): S20–S28.
31. Peers, C., Dallas, M.L., Boycott, H.E., et al. "Hypoxia and Neurodegeneration." *Annals of the New York Academy of Sciences* 1177 (October 2009): 169–77.
32. Wolf, S.A., Kronenberg, G., Lehmann, K., et al. "Cognitive and Physical Activity Differently Modulate Disease Progression in the Amyloid Precursor Protein (APP)-23 Model of Alzheimer's Disease." *Biological Psychiatry* 60 (2006): 1314–23.

33. Wiklund, L., Martijn, C., Miclescu, A., Semenas, E., Rubertsson, S., and Sharma H.S. "Central Nervous Tissue Damage after Hypoxia and Reperfusion in Conjunction with Cardiac Arrest and Cardiopulmonary Resuscitation: Mechanisms of Action and Possibilities for Mitigation." *International Review of Neurobiology* 102 (2012): 173–87.

34. Guo, Z. et al. "Head Injury and the Risk of AD in the MIRAGE Study." *Neurology* 54 (2000): 1316–23.

35. Sabedra, A.R., Kristan, J., Raina, K., Holm, M.B., Callaway, C.W., Guyette, F.X., Dezfulian, C., Doshi, A.A., and Rittenberger, J.C. "Neurocognitive Outcomes Following Successful Resuscitation from Cardiac Arrest." *Resuscitation* 90 (May 2015): 67–72.

36. Yaffe, K., Laffan, A.M., Harrison, S.L., Redline, S., Spira, A.P., Ensrud, K.E., Ancoli-Israel, S., and Stone, K.L. "Sleep-Disordered Breathing, Hypoxia, and Risk of Mild Cognitive Impairment and Dementia in Older Women." *JAMA* 306 (2011): 613–9.

37. Christianson, M.G. and Lo, D.C. "Differential Roles of Aβ Processing in Hypoxia-Induced Axonal Damage." *Neurobiology of Disease* 77 (May 2015): 94–105.

38. Liu, J., Killilea, D.W., and Ames, B.N. "Age-Associated Mitochondrial Oxidative Decay: Improvement of Carnitine Acetyltransferase Substrate Binding Affinity and Activity in Brain by Feeding Old Rats Acetyl-L-Carnitine or R-Alpha-Lipoic Acid." *Proceedings of the National Academy of Sciences (PNAS) of the United States of America* 99, no. 4 (February 19, 2002): 1876–81.

39. Hagen, T.M., Liu, J., Lykkesfeldt, J., Wehr, C.M., Ingersoll, R.T., Vinarsky, V., Bartholomew, J.C., and Ames, B.N. "Feeding Acetyl-L-Carnitine and Lipoic Acid to Old Rats Significantly Improves Metabolic Function while Decreasing Oxidative Stress." *PNAS* 99, no. 4 (February 19, 2002): 1870–5.

40. Carta, A. and Calvani, M. "Acetyl-L-Carnitine: A Drug Able to Slow the Progress of Alzheimer's Disease?" *Annals of the New York Academy of Sciences* 640 (1991): 228–32.

41. Li, S., Li, Q., Li, Y., Li, L., Tian, H., and Sun, X. "Acetyl-L-Carnitine in the Treatment of Peripheral Neuropathic Pain: A Systematic Review and Meta-analysis of Randomized Controlled Trials." *PLOS ONE* 10, no. 3 (March 9, 2015): e0119479.

42. Wang, S.M., Han, C., Lee, S.J., Patkar, A.A., Masand, P.S., and Pae, C.U. "A Review of Current Evidence for Acetyl-L-Carnitine in the Treatment of Depression." *Journal of Psychiatric Research* 53 (June 2014): 30–7.

43. Soczynska, J.K., Kennedy, S.H., Chow, C.S., Woldeyohannes, H.O., Konarski, J.Z., and McIntyre, R.S. "Acetyl-L-Carnitine and Alpha-Lipoic Acid: Possible Neuropathic Agents for Mood Disorders?" *Expert Opinion on Investigational Drugs* 17, no. 6 (June 2008): 827–43.

44. Sun, Q., Jia, N., Wang, W., Jin, H., Xu, J., and Hu, H. "Protective Effects of Astragaloside IV against Amyloid Beta1-42 Neurotoxicity by Inhibiting the Mitochondrial Permeability Transition Pore Opening." *PLOS ONE* 9, no. 6 (2014): e98866.

45. Horiuchi, M., Maezawa, I., Itoh, A., Wakayama, K., Jin, L.W., et al. "Amyloid Beta1-42 Oligomer Inhibits Myelin Sheet Formation in Vitro." *Neurobiology of Aging* 33 (2012): 499–509.

46. Chafekar, S.M., Hoozemans, J.J., Zwart, R., Baas, F., and Scheper, W. "Aβ 1-42 Induces Mild Endoplasmic Reticulum Stress in an Aggregation State-Dependent Manner." *Antioxidants and Redox Signaling* 9 (2007): 2245–54.

47. Yang, F., Lim, G.P., Begum, A.N., Ubeda, O.J., Simmons, M.R., Ambegaokar, S.S.; Chen, P.P., Kayed, R., Glabe, C.G., Frautschy, S.A., and Cole, G.M.

"Curcumin Inhibits Formation of Amyloid Beta Oligomers and Fibrils, Binds Plaques, and Reduces Amyloid in Vivo." *The Journal of Biological Chemistry* 280 (2005): 5892–5901.

48. Frautschy, S.A., Hu, W., Miller, S.A., Kim, P., Harris-White, M.E., and Cole, G.M. "Phenolic Anti-inflammatory Antioxidant Reversal of Aβ-Induced Cognitive Deficits and Neuropathology." *Neurobiology of Aging* 22 (2001): 993–1005.

49. Reddy, P.H., Manczak, M., Yin, X., Grady, M.C., Mitchell, A., Kandimalla, R., and Kuruva, C.S. "Protective Effects of a Natural Product, Curcumin, against Amyloid Beta-Induced Mitochondrial and Synaptic Toxicities in Alzheimer's Disease." *Journal of Investigative Medicine* (August 12, 2016). pii: jim-2016-000240.

50. Baum, L., Lam, C.W., Cheung, S.K., Kwok, T., Lui, V., and Tsoh, J., et al. "Six-Month Randomized, Placebo-Controlled, Double-Blind, Pilot Clinical Trial of Curcumin in Patients with Alzheimer's Disease." *Journal of Clinical Psychopharmacology* 28 (2008): 110–3.

51. Liu, Z-J., Li, Z-H., Liu, L., Tang, WX., Wang, Y., Dong, M-R., and Xiao, C. "Curcumin Attenuates Beta-Amyloid-Induced Neuroinflammation via Activation of Peroxisome Proliferator-Activated Receptor-Gamma Function in a Rat Model of Alzheimer's Disease." *Frontiers in Pharmacology* 7 (2016): 261.

52. Levites, Y., Amit, T., Mandel, S., and Youdim, M.B. "Neuroprotection and Neurorescue against Aβ Toxicity and PKC-Dependent Release of Nonamyloidogenic Soluble Precursor Protein by Green Tea Polyphenol (–)-Epigallocatechin-3-Gallate. *The FASEB Journal* 17 (2003): 952–54.

53. Rezai-Zadeh, K., Shytle, D., Sun, N., Mori, T., Hou, H., et al. "Green Tea Epigallocatechin-3 Gallate (EGCG)

Modulates Amyloid Precursor Protein Cleavage and Reduces Cerebral Amyloidosis in Alzheimer Transgenic Mice. *Journal of Neuroscience* 25 (2005): 8807–14.

54. Noguchi-Shinohara, M., Yuki, S., Dohmoto, C., Ikeda, Y., Samuraki, M., Iwasa, K., Yokogawa, M., Asai, K., Komai, K., Nakamura, H., and Yamada, M. "Consumption of Green Tea, but Not Black Tea or Coffee, Is Associated with Reduced Risk of Cognitive Decline." *PLOS ONE* 9, no. 5 (May 14, 2014).

55. Zhang, S.Q., Sawmiller, D., Li, S., Rezai-Zadeh, K., Hou, H., Zhou, S., Shytle, D., Giunta, B., Fernandez, F., Mori, T., and Tan, J. "Octyl Gallate Markedly Promotes Anti-amyloidogenic Processing of APP through Estrogen Receptor-Mediated ADAM10 Activation." *PLOS ONE* 8, no. 8 (August 15, 2013).

56. Caprilla, H., Vintdeux, V., and Zhao, H. "Resveratrol Mitigates Lipopolysaccharide- and Aβ-mediated Microglial Inflammation by Inhibiting the TLR4/NF-κB/STAT Signaling Cascade." *Journal of Neurochemistry* 120, no. 3 (February 2012): 461–72.

57. Zhang, F.,; Liu, J., and Shi, J.S. "Anti-inflammatory Activities of Resveratrol in the Brain: Role of Resveratrol in Microglial Activation." *European Journal of Pharmacology* 636 (2010): 1–7.

58. Ghobeh, M., Ahmadian, S., Meratan, A.A., Ebrahim-Habibi, A., Ghasemi, A., Shafizadeh, M., and Nemat-Gorgani. M. "Interaction of a Beta (25-35) Fibrillation Products with Mitochondria: Effect of Small-Molecule Natural Products." *Biopolymers* 102, no. 6 (November 2014): 473–86.

59. Sabogal-Guáqueta, M., Muñoz-Manco, J., Ramírez-Pineda, J., Marisol, L., Edison, O., and Cardona Gómez, P. "The Flavonoid Quercetin Ameliorates Alzheimer's Disease Pathology and Protects Cognitive and Emotional Function in Aged Triple Transgenic Alzheimer's Disease Model Mice." *Neuropharmacology* 93 (June

2015): 134–45.

60. Ono, K., Yoshiike, Y., Takashima, A., Hasegawa, K., Naiki, H., and Yamada, M. "Potent Antiamyloidogenic and Fibril-Destabilizing Effects of Polyphenols in Vitro: Implications for the Prevention and Therapeutics of Alzheimer's disease." *Journal of Neurochemistry* 87 (2003): 172–81.
61. Choi, Y., Kim, T.D., Paik, S.R., Jeong, K., and Jung, S. "Molecular Simulations for Antiamyloidogenic Effect of Flavonoid Myricetin Exerted against Alzheimer's β-Amyloid Fibrils Formation." *Bulletin of the Korean Chemistry Society* 29 (2008): 1505–9.
62. Lee, M., McGeer, E.G., and McGeer, P.L. "Quercetin, Not Caffeine, Is a Major Protective Component in Coffee." *Neurobiology of Aging* 46 (October 2016): 113–23.
63. Wang, J.B., Wang, Y.M., and Zeng, C.M. "Quercetin Inhibits Amyloid Fibrillation of Bovine Insulin and Destabilizes Preformed Fibrils." *Biochemical and Biophysical Research Communications* 415, no. 4 (December 2, 2011): 675–9.
64. Jiménez-Aliaga, K., Bermejo-Bescós, P., Benedí, J., and Martín-Aragón, S. "Quercetin and Rutin Exhibit Anti-amyloidogenic and Fibril-Disaggregating Effects in Vitro and Potent Antioxidant Antioxidant Activity in APPswe Cells." *Life Sciences* 89, no. 25-26 (December 19, 2011): 939–45.
65. Ndaqitimana, A., Wang, X., Pan, G., et al. "A Review on Indole Alkaloids Isolated from Uncaria Rhynchophylla and Their Pharmacological Studies." *Fitoterapia* 86 (April 2013): 35–47.
66. Tan, S., Wood, M., and Maher, P. "Oxidative Stress Induces a Form of Programmed Cell Death with Characteristics of Both Apoptosis and Necrosis in Neuronal Cells." *Journal of Neurochemistry* 71 (1998): 95.

67. Qi, W., Yue, S.J., Sun, J.H., et al. "Alkaloids from the Hook-Bearing Branch of Uncaria Rhynchophylla and Their Neuroprotective Effects against Glutamate-Induced HT22 Cell Death." *Journal of Asian Natural Products Research* 16, no. 8 (2014): 876–83.

68. Rackayova, V., Cudalbu, C., Pouwels, P.J., and Braissant, O. "Creatine in the Central Nervous System: From Magnetic Resonance Spectroscopy to Creatine Deficiencies." *Analytical Biochemistry* (November 10, 2016). pii: S0003-2697(16)30389-X.

69. Perasso, L., Spallarossa, P., Gandolfo, C., Ruggeri, P., Balestrino M. "Therapeutic Use of Creatine in Brain or Heart Ischemia: Available Data and Future Perspectives." *Medicinal Research Reviews* 33, no. 2 (March 2013): 336–63.

70. Attia, A., Ahmed, H., Gadelkarim, M., Morsi, M., Awad, K., Elnenny, M., Ghanem, E., El-Jafaary, S., and Negida, A. "Meta-analysis of Creatine for Neuroprotection against Parkinson's Disease." *CNS and Neurological Disorders - Drug Targets* (November 4, 2016).

71. Pastula, D.M., Moore, D.H., and Bedlack, R.S. "Creatine for Amyotrophic Lateral Sclerosis/Motor Neuron Disease." *The Cochrane Database of Systematic Reviews* (December 12, 2012). 12:CD005225.

72. Kumar, A. and Singh, A. "A Review on Mitochondrial Restorative Mechanism of Antioxidants in Alzheimer's Disease and Other Neurological Conditions." *Frontiers in Pharmacology* 6 (September 24, 2015): 206.

73. Zervou, S., Whittington, H.J., Russell, A.J., and Lygate, C.A. "Augmentation of Creatine in the Heart." *Mini-Reviews in Medicinal Chemistry* 16, no. 1 (2016): 19–28. Review.

Chapter 8

Mitochondrial Dysfunction and Arthritis

As we age, we notice more stiffness and, in some cases, aching or pain in our joints. You may think this is just normal wear and tear on the body or the result of some previous injury. Generally, your first thought isn't what effect your mitochondria may have on this process. However, you'll find out in this chapter what a significant driver your mitochondria are in the joint inflammation process and, more importantly, what you can do to help your joints.

Arthritis is a disease characterized by inflammation—the body's natural response to injury—and/or degeneration, especially of the joints and other tissue structures. The disease most often presents with joint tenderness, pain, swelling, stiffness, and reductions in joint mobility and range of movement. Typically, arthritis pain and degeneration of joints persist over many years, and these slowly increase over time. However, certain events, such as trauma or infection(s), can cause sudden increases in the severity of arthritic signs and symptoms.

Some Common Types of Arthritis

Osteoarthritis (OA) is the most common form of arthritis, and it can affect both large and small joints of the body, including the hands, feet, back, hips, or knees. Osteoarthritis is a cartilage and joint disease related to age and characterized by a reduction in the number of chondrocytes (cartilage cells), loss of the extracellular matrix, and synovial inflammation (1, 2). The disease is essentially caused by daily wear and tear on the joints and their protective coverings; however, OA can also occur as a result of injury or infection. Osteoarthritis begins in the cartilage and eventually causes the cartilage between the opposing bones in the joint to erode, thus narrowing the gap between these bones, which leads to pain and reduced joint mobility. Osteoarthritis typically affects the weight-bearing joints, such as those in the back, spine, and pelvis.

Osteoarthritis usually afflicts the elderly. More than 30 percent of women have some degree of OA by age sixty-five. Risk factors for OA include prior joint trauma, obesity, infections, and a sedentary lifestyle.

Rheumatoid arthritis is an autoimmune disease in which the body's immune system attacks the joints, mostly the joint lining and cartilage, as well as other body tissues. This eventually results in erosion of opposing bones in the joints along with swelling, often in the fingers, wrists, knees, elbows, and ankles. The disease is usually symmetrical and can lead to severe deformity in a few years if untreated.

Rheumatoid arthritis occurs mostly in people age twenty and older and, in contrast to osteoarthritis, this condition is a systemic or system-wide autoimmune disease affecting many other tissues and organs.

Possible Causes of Arthritis and Treatment of Symptoms

The causes of various common forms of arthritis are generally unknown, and most research indicates there's no one single cause for the various types of arthritis. Multiple factors are involved that are probably different in each patient and type of arthritis.

For the most part, treatments for rheumatic diseases have depended on addressing symptoms, such as alleviating pain and inflammation. These treatments aren't directed at possible causes of the disease.

Mitochondrial Dysfunction, Chondrocytes, and Autophagy

An important property of tissues involved in arthritis is that energy functions inside cells—in particular, the cells of the joints—are dysfunctional. This has been found in the most common forms of arthritis, osteoarthritis and rheumatoid arthritis (1, 2). These conditions result in increased oxidative stress and the excess production of ROS, the free-radical oxygen species that can directly damage cellular membranes (1).

Chondrocytes and the cartilage matrix are important components of the cartilage structure. Chondrocytes are highly specialized cells involved in synthesizing collagen and proteoglycans as well as maintaining cartilage homeostasis. Chondrocytes contain mitochondria that are susceptible to ROS, lipid oxidation, and hypoxia. Well-recognized cartilage diseases attributable to the extracellular matrix (ECM) degeneration that affects normal chondrocyte function include osteoarthritis and articular cartilage injury (1, 2). Importantly, chondrocyte death suppresses synthesis of the cartilage matrix, resulting in delayed tissue remodeling and poor recovery (1, 2). Cartilage tissue undergoes continual internal remodeling as the cells replace cartilage matrix molecules lost through degradation. Normal matrix turnover depends on the ability of chondrocytes to detect changes in the matrix and then build new molecules. Bearing weight on the joints stimulates the activity of the chondrocytes. Diminished use of the joints leads to alterations in the matrix and, eventually, tissue structure and mechanical properties are lost. Using the joint stimulates the chondrocytes to repair the cartilage matrix (1, 2).

Nitric Oxide

Nitric oxide (NO) is a messenger implicated in the destruction and inflammation of joint tissues. There are three enzymes that create NO, but the destructive one in joints is inducible nitric oxide synthase (iNOS), which is expressed for longer periods of time. Its overexpression is caused by inflammatory cytokines, which gradually increase with age (3).

Cartilage and synovial membranes in patients with rheumatoid arthritis and osteoarthritis have high levels of NO. NO is known to modulate various cellular pathways, inhibit the activity of the mitochondrial respiratory chain (MRC) of chondrocytes, and induce the generation of reactive oxygen species (ROS) and cell death in multiple cell types (3).

Autophagy

Autophagy is a cellular degradation pathway of organelles and proteins activated by conditions of cellular stress, including hypoxia, ROS, endoplasmic reticulum (ER) stress, and microbial infection. Autophagy is essential for health and survival. As a protective mechanism, autophagy safeguards organisms against normal and pathological aging by regulating the turnover of dysfunctional organelles and proteins. In cellular processes in which nutrients are low and energy demands are high (such as growth or differentiation), autophagy is activated to maintain energy metabolism (4).

In the joints, autophagy aids chondrocyte cells by sending signals to direct their synthesis of cartilage components in areas where joint cartilage is eroding (4).

Cumulative oxidative stress and excess ROS, as seen in arthritis and other chronic diseases, can damage mitochondria. At a higher systemic level, oxidative stress causes fatigue and nerve cell damage at synapses, the communication regions between nerve cells. Thus, changes in mitochondria can influence both the onset and severity of arthritis and the fatigue and pain that accompany it (5, 6).

Hypoxia

Hypoxia has been implicated in driving the pro-inflammatory response, although its role in relation to mitochondrial dysfunction in inflammatory arthritis is less understood. Mitochondria require oxygen to power the electron transport chain. It's understood that in vivo oxygen levels must be maintained within certain strict limits to avoid mitochondrial mutations and ROS (7, 8).

Hypoxia results in reduced hemoglobin and ferritin levels. Hypoxia-induced iron deficiency may cause elevated NO levels through another physiological mechanism by causing anemic hypoxia—subnormal oxygen content in the blood. Hypoxia is known to cause a stimulation of NO production through iNOS, the enzyme that causes excess NO production. This causes a loss of mitochondrial membrane potential and fragmentation of mitochondrial DNA. In addition to contributing to the breakdown of the extracellular matrix, NO also causes apoptosis by a mitochondria-dependent mechanism (8).

Hypoxia induces a wide spectrum of alterations in mitochondrial structure, dynamics, and genome stability, resulting in reduced mitochondrial respiration, excessive production of ROS, loss of ATP, increased oxidative damage, and the accumulation of mtDNA mutations (8).

Several processes drive the hypoxia that causes mitochondrial dysfunction and the resulting damage to chondrocytes and the cartilage matrix structure (8).

First (and the one most easily understood) is the effect of sitting most of the day. Biomechanically, the hamstring muscles (there are three: biceps femoris, semitendinosus, and semimembranosus) attach to our ischial tuberosity proximally and to the femur, tibia, and fibula distally, which means they attach to your pelvis, your thigh, and your lower leg (see the figure below). We usually think of their action, collectively, as knee flexors—as in, if you need to bring your heel closer to the back of your thigh, they're on the job. But we commonly neglect to pay attention to their other action, which is to posteriorly tilt the pelvis—as in, if you want to flatten your lower back and tuck your pelvis under, they perform that function, too.

Many of us sit on our sacrums instead of on our ischial tuberosities (the fictional "sit bones"), which means we're sitting on a posteriorly tilted pelvis all day long. Your lumbar discs and sacroiliac ligaments greatly dislike this, and so do your hamstrings. Sitting like this means the hamstrings are constantly contracted, which means the flow of blood through the vascular network is diminished. This limits the delivery of nutrients and oxygen to the joints, causing hypoxia.

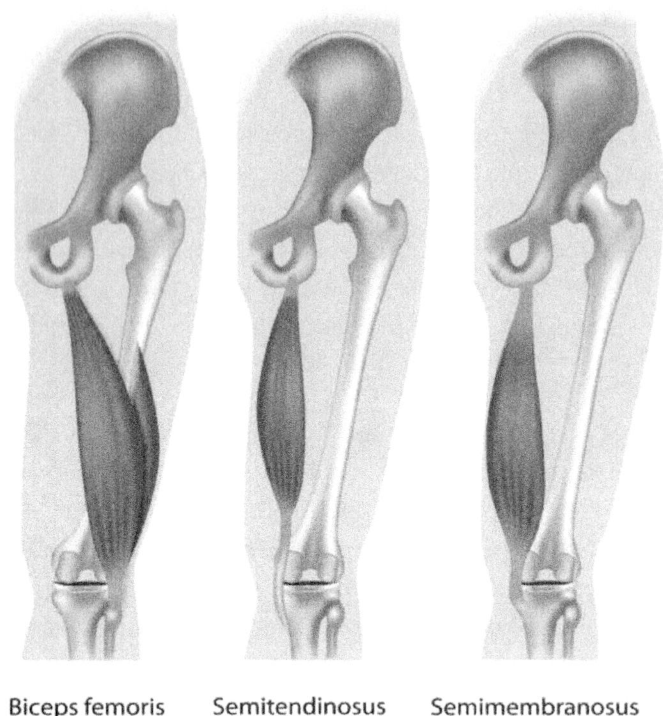

Figure 6. The three hamstring muscles
Credit: Shutterstock

Another cause of hypoxia is the narrowing of the arteries as a result of plaque buildup, which reduces blood flow to the joints by increasing intima media thickness resulting in lower flow-mediated dilation. In patients who have early-stage rheumatoid arthritis, flow-mediated dilation was reduced about 50 percent compared to control subjects (9).

Plaque buildup in the heart (athersclerosis) also reduces the delivery of nutrients and oxygen to the joints. This hypoxia in the joints drives chronic inflammation (10).

Other forms of lifestyle-related hypoxia may also be present and have similar effects on the mitochondria and the joints. For example, diffusional hypoxia results from damaged pulmonary membranes and impaired lung function, which happens in chronic obstructive pulmonary disease (COPD, often caused by smoking). Therefore, lung diseases may result in elevated NO levels by causing chronic hypoxia (11).

Impaired lung function may also be pharmacologically induced with the use of common medications such as beta blockers. These drugs have the potential to promote bronchospasm and bronchoconstriction. This class of drugs is often used to treat hypertension, cardiac arrhythmia, chronic angina pectoris, and other conditions (12).

As for natural treatments, some nutrients have been shown to enhance lung function. The most notable of these is N-acetylcysteine, commonly referred to as NAC. NAC, found in cruciferous vegetables, such as broccoli and cabbage, is a dietary source of L-cysteine and is one of three amino acids the body uses to assemble glutathione, the ubiquitous sulfhydryl antioxidant inside cells that's rapidly depleted during episodes of oxidative stress. However, NAC has direct antioxidant properties that cannot be explained as simply being a dietary source of L-cysteine (13).

Therapeutic Interventions to Improve Joint Health

As we've shown, the number and well-being of mitochondria are vital to proper joint function and lack of inflammation. Many of the available interventions to improve mitochondrial health focus on mitochondrial biogenesis and reduced oxidative stress. The combination of diverse approaches—spanning dietary alterations, natural supplements, and lifestyle changes—will likely produce optimal results for improving neurological outcomes.

Hydroxytyrosol

Extra virgin olive oil (EVOO) is a bioactive food found in the Mediterranean diet and is associated with a reduced risk of pathologies related to immune/inflammatory diseases (14). Extra virgin olive oil is also the only type of olive oil that contains the polyphenol hydroxytyrosol (14).

Cell culture studies show that hydroxytyrosol extends chronological lifespan (CLS) in fibroblasts by enhancing manganese superoxide dismutase (MnSOD) activity. CLS is defined as the time in which normal cells retain their capacity to be in a proliferative, actively dividing cell cycle. MnSOD helps suppress age-associated accumulation of mitochondrial ROS damage. MnSOD is the only type of SOD found exclusively in mitochondria (15).

Hormesis is the process of creating tiny amounts of cellular damage that causes a disproportionate production of endogenous antioxidant enzymes (16).

Hydroxytyrosol is dismutated, or converted into superoxide radicals, which then causes a small amount of mitochondrial DNA damage that provokes an overwhelming response by the mitochondrial genes to produce higher than normal amounts of MnSOD (15). As cells age, their mitochondrial MnSOD levels decline and mitochondrial free-radical (ROS) damage accumulates. Adding hydroxytyrosol to human cell cultures increases their MnSOD levels and their ability to continually divide, increasing their CLS (15). This effect at the mitochondrial level helps explain why EVOO has multiple health benefits.

Methyl Sulfonyl Methane

Methyl sulfonyl methane (MSM) has been shown to improve

symptoms of pain and physical function in arthritis without major adverse events. MSM has been used to treat osteoarthritis of the knee in clinical trials, where it resulted in better pain improvement scores than placebo (17).

MSM is naturally present in the human body in small amounts; it can be found in the cerebrospinal fluid (18).

MSM works through growth hormone (GH)-related pathways. MSM and GH, separately or in combination, activate growth-hormone signaling via the common STAT5b pathway. Six osteogenic marker genes that build bone (ALP, ON, OCN, BSP, OSX, and Runx2) are activated by MSM. MSM increases the mineralization of mesenchymal stem cells (MSCs) and promotes osteogenic differentiation of MSCs into bone cells through the activation of the STAT5b gene (19).

Astragalus

Astragalus root powder and its extracts are key ingredients in traditional Chinese herbal decoctions used to strengthen the lungs. For thousands of years, astragalus has been recommended for invigoration, health preservation, and reduction of fatigue, and it is held in high esteem. It's also believed to have anti-aging properties (20, 21).

Modern research has shown that two of its ingredients, Astragaloside IV and cycloastragenol, increase critically shortened leukocyte telomeres in humans (21). Astragaloside IV is used as a drug to treat cardiovascular diseases in China, and its proponents believe that the treatment can reverse atheroscle (22). When Astragaloside IV is added to cultured human endothelial cells from diabetic patients, it increases telomere length in cells with critically shortened telomeres (23).

Resveratrol

Resveratrol, a phytoalexin present in grape skin and red wine, exerts a variety of actions to reduce superoxides, suppress carcinogenesis and angiogenesis, prevent diabetes mellitus, inhibit inflammation, and prolong the lifespan (5). Furthermore, resveratrol decreases plaque formation associated with neurodegenerative diseases such as Alzheimer's disease and Huntington's disease (6).

Lipid Replacement Therapy

Because arthritis patients have mitochondrial and other membrane impairments, supplementing the diet with the right dietary lipids is an especially attractive, all-natural approach to repair cell membrane damage and reverse the effects of damage caused by excess ROS. There are three specific types of membrane lipids—phospho-lipids, glycolipids, and cholesterol—that can repair mitochondria and cell membranes to make them more functional (26).

Lipid replacement therapy is an effective way to reduce the effects of oxidative damage to the cells in our joints, nerves, and other tissues and restore mitochondrial function. Restoring mitochondrial function and repairing cellular membranes reduces fatigue and restores cellular function (26). Cardiolipin is a phospholipid necessary for mitochondrial electron transport in the inner mitochondrial membrane, and its levels decline with age. Dietary lipids are a building block needed to replace lost cardiolipin in mitochondria (26).

References:

1. Kapoor, M. and Mahomed,, N. *Osteoarthritis. Pathogenesis, Diagnosis, Available Treatments, Drug Safety, Regenerative and Precision Medicine* (New York: Springer Publishing, 2015).

2. "Rheumatoid Arthritis: In Depth," NIH Center for Complementary and Alternative Health, NCCIH pub. no. D441, last updated July 2013. https://nccih.nih.gov/health/RA/getthefacts.htm.

3. Rahmati, M., Mobasheri, A., and Mozafari, M. "Inflammatory Mediators in Osteoarthritis: A Critical Review of the State-of-the-Art, Current Prospects, and Future Challenges." *Bone* 85 (April 2016): 81–90. Review.

4. Rockel, J.S. and Kapoor, M. "Autophagy: Controlling Cell Fate in Rheumatic Diseases."

   Nature Reviews Rheumatology 12, no. 9 (September 2016): 517–31. Review.

5. Mårtensson, C.U., Doan, K.N., and Becker, T. "Effects of Lipids on Mitochondrial Functions." *Biochimica et Biophysica Acta* 1862, no. 1 (January 2017): 102–13.

6. Aufschnaiter, A., Kohler, V., Diessl, J., Peselj, C., Carmona-Gutierrez, D., Keller, W., and Büttner, S. "Mitochondrial Lipids in Neurodegeneration." *Cell and Tissue Research* 367, no. 1 (January 2017): 125–40.

7. Nicolson, G., Ellithorpe, R., et al. "Lipid Replacement Therapy with a Glycophospholipid-Antioxidant-Vitamin Formulation Significantly Reduces Fatigue within One Week." *Journal of the American Nutraceutical Association*. 13, no. 1 (2010): 163–70.

8. Lane, R.S., Fu, Y., Matsuzaki, S., Kinter, M., Humphries, K.M., and Griffin, T.M. "Mitochondrial Respiration and Redox Coupling in Articular Chondrocytes." *Arthritis Research and Therapy* 17 (March 10, 2015): 54.

9. Chatterjee Adhikari, M., Guin, A., Chakraborty, S., Sinhamahapatra, P., and Ghosh, A. "Subclinical Atherosclerosis and Endothelial Dysfunction in Patients with Early Rheumatoid Arthritis as Evidenced by Measurement of Carotid-Media Thickness and Flow-Mediated Vasodilation: An Observational Study." *Seminars in Arthritis and Rheumatism* 41, no. 5 (April 2012): 669–75.

10. Södergren, A., Karp, K., Bengtsson, C., Möller, B., Rantapää-Dahlqvist, S., and Wållberg-Jonsson, S. "The Extent of Subclinical Atherosclerosis Is Partially Predicted by the Inflammatory Load: A Prospective Study over 5 Years in Patients with Rheumatoid Arthritis and Matched Controls." *The Journal of Rheumatology* 42, no. 6 (June 2015): 935–42.

11. Tao, H.,. Luo, W., Pei, H., et al. "Expression and Significance of Hypoxia-Inducible Factor-1a in Patients with Chronic Obstructive Pulmonary Disease and Smokers with Normal Lung Function." *Chinese Journal of Cellular and Molecular Immunology* 30, no. 8 (August 2014): 852–5.

12. Dollery, C.T., Paterson, J.W., and Conolly, M.E. "Clinical Pharmacology of Beta-Receptor-Blocking Drugs." *Clinical Pharmacology and Therapeutics* 10, no. 6 (November-December 1969): 765–99. Review.

13. Elbini Dhouib, I., Jallouli, M., Annabi, A., Gharbi, N., Elfazaa, S., and Lasram, M.M. "A Mini-Review on N-acetylcysteine: An Old Drug with New Approaches." *Life Sciences* 151 (April 15, 2016): 359–63.

14. Aparicio-Soto, M., Sánchez-Hidalgo, M., Rosillo, M.Á., Castejón, M.L., and Alarcón-de-la-Lastra, C. "Extra Virgin Olive Oil: A Key Functional Food for Prevention of Immune-Inflammatory Diseases." *Food and Function* 7, no. 11 (November 9, 2016): 4492–505. Review.

15. Sarsour, E.H., Kumar, M.G., Kalen, A.L., Goswami, M., Buettner, G.R., and Goswami P.C. "MnSOD Activity

Regulates Hydroxytyrosol-Induced Extension of Chronological Lifespan." *Age (Dordrecht, Netherlands)* 34, no. 1 (February 2012): 95–109.

16. Kendig, E.L., Le, H.H., and Belcher, S.M. "Defining Hormesis: Evaluation of a Complex Concentration Response Phenomenon." *International Journal of Toxicology* 29, no. 3 (May-June 2010): 235–46.
17. Brien, S., Prescott, P., and Lewith, G. "Meta-Analysis of Related Nutritional Supplements Dimethyl Sulfoxide and Methylsulfonylmethane in the Treatment of Osteoarthritis of the Knee." *Evidence-Based Complementary and Alternative Medicine* (*eCAM*) 2011, article ID no. 528403 (2011).
18. Engelke, U.F., Tangerman, A., Willemsen, M.A., Moskau, D., Loss, S., Mudd, S.H., and Wevers, R.A. "Dimethyl Sulfone in Human Cerebrospinal Fluid and Blood Plasma Confirmed by One-Dimensional (1) H and Two-Dimensional (1) H-(13) C NMR." *NMR in Biomedicine* 18 (2005): 331–6.
19. Joung, Y.H., Lim, E.J., Darvin, P., Chung, S.C., et al. "MSM Enhances GH Signaling via the Jak2/STAT5b Pathway in Osteoblast-Like Cells and Osteoblast Differentiation through the Activation of STAT5b in MSCs." *PLOS ONE* 7, no. 10 (2012): e47477.
20. Shahzad, M., Shabbir, A., Wojcikowski, K., Wohlmuth, H., and Gobe, G.C. "The Antioxidant Effects of Radix Astragali (Astragali Membranaceus and Related Species in Protecting Tissues from Injury and Disease." *Current Drug Targets* 17, no. 12 (2016): 1331–40. Review.
21. Shen, C.Y., Jiang, J.G., Yang, L., Wang, D.W., and Zhu, W. "Anti-ageing Active Ingredients from Herbs and Nutraceuticals Used in Traditional Chinese Medicine: Pharmacological Mechanisms and Implications for Drug Discovery." *British Journal of Pharmacology* (September 23, 2016). doi: 10.1111/bph.13631. Review.
22. Miller, A.L. "Botanical Influences on Cardiovascular

Disease." *Alternative Medicine Review* 3 (1998): 422–31.

23. Li, M., Yu, L., She, T., Gan, Y., et al. "Astragaloside IV Attenuates Toll-Like Receptor 4 Expression via NF-kB Pathway under High-Glucose Condition in Mesenchymal Stem Cells." *European Journal of Pharmacology* 696, no. 1–3 (December 5, 2012): 203–9.
24. Elliott, P.J. and Jirousek, M. "Sirtuins: Novel Targets for Metabolic Disease." *Current Opinion in Investigational Drugs* 9, no. 4 (April 2008): 371–8.
25. Pinto, J.T., Xu, H., Chen, H.L., Beal, M.F., Gibson, G.E., and Karuppagounder, S.S. "Dietary Supplementation with Resveratrol Reduces Plaque Pathology in a Transgenic model of Alzheimer's Disease. *Neurochemistry International* 54, no. 2 (February 2009): 111–8.
26. Nicholson, G.L. and Ash, M.E. "Lipid Replacement Therapy: A Natural Medicine Approach to Replacing Damaged Lipids in Cellular Membranes and Organelles and Restoring Function." *Biochemica et Biophysica Acta* 1838, no. 6 (June 2014): 1657–79.

Chapter 9

The Aging Effects of Mitochondrial Dysfunctions on the Skin, Eyes, and Muscles

"How old would you be if you didn't know how old you was?"
--Satchel Paige (1906–1982)

The most noticeable effects of the aging process are the changes to our skin. Changes in our skin's appearance are even more noticeable to people who haven't seen us in a long time. However, we're constantly reminded of the changes in our muscles and our vision. The aging process that affects our skin, muscles, and eyes is driven, in large part, by mitochondrial dysfunction, the same process that drives the aging of our internal organs and tissues.

While it may be easier to overlook some of the symptoms of other mitochondrial age-related diseases, such as diabetes or cardiovascular illness, we're constantly reminded that we're getting older by the stiffness in our muscles, our lack of flexibility, and our wrinkles and sagging skin. Whether we try to read close up or far away, our vision isn't as acute as it was when we were younger.

Advanced Glycation End Products

Before we can begin our discussion of how mitochondria affect our skin, muscles, and eyes, we need to discuss a major factor that drives mitochondrial dysfunction. We've known for several decades that the process of glycosylation, now called glycation, is one of the most important causes of aging—equally as important as the process of protein oxidation and cell membrane lipid peroxidation (where ROS oxidize lipids to become rancid oils). Glycation occurs during our daily metabolism, which involves sugar molecules—glucose and fructose, aldehydes, and ketones—cross-linking with amino groups on proteins in a process called nonenzymatic glycation. This reaction, also known as the Maillard reaction, acts like a pair of handcuffs that cross-links proteins in a random, unnatural way. This process is distinct from programmed enzymatic glycation, which is part of normal metabolism and cellular function. Glycation occurs when glucose is enzymatically split open by enzymes in the cytosol (the liquid inside cells) to generate energy. The split glucose ring has an aldehyde exposed, and this aldehyde can cross-link adjacent proteins.

The body's repair enzymes can reverse early glycation end products, but once they form advanced glycation end products (AGEs), these repair enzymes have far more trouble removing them from proteins, including from DNA. This random glycation process results in glycated protein-aldehyde adducts (a product created by the addition of molecules) that accumulate with age. These compounds can, and usually do, react with other proteins, resulting in irreversible bonding between the two. Beta amyloid is an example of a cross-linked protein aggregate that is a byproduct of AGE formation and, most notably, accumulates in the brain, a contributing factor in Alzheimer's disease (1).

Glycation-affected molecules can range from collagen and elastin to enzymes and immune proteins. Facilitators during cross-linking are the carbonyl groups, which act like glue, fixing the two proteins together. Carbonyls are chemical groups that are formed as a result of a sugar (or an aldehyde or a ketone or a free radical) reacting with amino acids on a protein. Carbonyls cause not only protein-to-protein cross-linking but also protein-to-DNA or protein-to-lipid cross-linking, all of which are equally damaging to the organism (1).

Carbonyls are one source of protein oxidation and provide the necessary anchor where glycation can build up, using the carbonyl as an anchor. Cross-linking results in the formation of large insoluble aggregates of damaged proteins in the tissues. AGEs may then go on to interact with free radicals and cause further tissue injury through chronic oxidation. Although a steady rate of AGE formation occurs during normal aging (starting after the age of twenty), AGE formation is accelerated during hyperglycemic states such as diabetes. As we mentioned, AGE bonds are difficult to break. Although there has been intensive research into how to accomplish this, nothing has been successful. AGEs are extremely prevalent in collagen, tissues of the eyes, and in endothelial cells (2).

Once formed, AGEs inhibit cellular transport processes, such as the removal of waste or intake of nutrients. AGEs stimulate mitochondria to generate more free radicals (such as superoxide and nitric oxide) and activate pro-inflammatory cytokines tumor necrosis factor alpha (TNF-alpha) and interleukin 6. In addition, some AGEs are immunogenic, causing age-related autoimmunity, or mutagenic, increasing the risk of cancer. AGEs also increase adhesion molecule activity, reduce protein-degradation rates by inhibiting protease activity, and reduce cell proliferation. All these AGE-caused disruptions of normal cellular function contribute to degenerative-disease development. Also, AGEs stimulate apoptosis, resulting in excessive cell death that further contributes to the risk of degeneration. Some AGEs upregulate chronic inflammatory cytokines.

At the clinical level, cross-linking contributes significantly to diabetic complications, lowers immune function, and increases the risk of cancer, atherosclerosis, and hypertension. The presence of AGEs increases Alzheimer's disease risk via accelerated beta-amyloid formation, which, as previously mentioned, is a cross-linked protein aggregate closely associated with AGE formation (2).

Cataracts, kidney damage, skin aging, and other age-related diseases are also strongly associated with the presence of AGEs. Both beta amyloid and AGEs increase oxidative stress, cause irritation to the surrounding cells, and can be considered as independent free-radical-generating systems (1).

We'll now discuss how AGEs and dysfunctional mitochondria interact to drive the aging process in our skin, muscles, and eyes.

The Skin

The skin is the largest organ in the body, covering an area on average of about 1.49 square meters. Its main function is to protect the body from insults originating in the external world. Another function is to provide sensory information about temperature, touch, pain, and vibration signals. Finally, the skin is also a secretory organ that eliminates toxins through perspiration.

The skin is a lipophilic barrier that protects against the absorption of water, which would otherwise dissolve our intracellular components.

The epidermis is the outermost layer of the skin and protects the skin against insults or injuries. It's often difficult for nutrients and pharmaceutical agents to penetrate the epidermis and reach the inner parts of the skin. Scientists have invested a great deal of research into finding effective ways of delivering nutrients where they can be effective—which is deep into the skin.

Liposomes are one such delivery method. These make use of tiny particles surrounded by lipid molecules that can penetrate the outer layers of the skin, carrying the nutrient or other chemical with them.

The main constituents of the dermis (the inner layer of skin) are collagen and elastin. The collagen and elastin infrastructure keeps the skin resilient, elastic, and firm. Collagen fibers are strong protein molecules that form a scaffolding to support other skin structures. Elastin fibers also form a mesh, or net, allowing the skin to maintain its elasticity and firmness. These two types of molecules can be considerably weakened by age-related processes due to glycation and oxidation. Photodamage, i.e. damage caused by the ultraviolet (UV) radiation from the sun, is also a major cause of collagen and elastin destruction.

Apart from the collagen/elastin network, other components of the dermis include fibroblasts, fatty tissues, nerves, blood vessels, sweat glands, hair follicles, and sensory structures. A major component of the skin is the intracellular matrix. This is formed by proteoglycans, a mixture of protein/carbohydrate molecules surrounded by water. The primary functions of the intracellular matrix are to facilitate the nutrition of the various skin components and to protect the skin from external insults.

The fluid nature of the intracellular matrix gives skin its plump, youthful, and supple appearance. Within the matrix, moisture is retained by substances such as hyaluronic acid and sodium pyrrolidone carboxylate (NaPCA), which are natural moisturizers. NaPCA levels decrease with age, causing a progressive loss of moisture in the skin. Hyaluronic acid can retain two hundred water molecules around each of its own molecules, and its production also declines with age. However, skin aging is a complicated process, involving several other interrelated factors.

How Aging Affects the Skin

Photoaging plays a major role in skin aging by causing direct damage to the skin followed by an exaggerated immune repair reaction, resulting in collagen and elastin degeneration, dilation of the small vascular vessels, deposition of abnormal elastin and collagen, and increased skin pigmentation. Other extrinsic aging factors include oxidative stress, pollution, and miscellaneous environmental insults. Intrinsic skin aging, on the other hand, is due to genetic factors and endogenous influences such as auto-oxidation, glycation, protein oxidation, and lipid peroxidation.

Free radicals are the main culprits commonly implicated in skin aging. Oxygen and nitrogen free radicals are produced by cigarette smoke, environmental pollutants, poor nutrition, pesticides, endocrine-disrupting chemicals (EDCs), and other food contaminants. UVA and UVB sunlight exposure is still the leading cause of free-radical production in the skin. UV radiation interacts with the DNA inside skin cells, causing damage that results in faulty DNA replication and imperfect DNA transcription. Defective protein formation from DNA mutations is perpetuated during subsequent cell divisions.

Chromophore skin cells are specialized cells that contain mitochondria that absorb a considerable amount of UV radiation. With age, chromophore mitochondrial activities diminish, resulting in more UV radiation reaching the inner parts of the skin. This causes an increase in electron transport, nitric oxide release from adenosine monophosphate, and increased blood flow. Increased internal ROS generation and activation of diverse signaling pathways also occur with chondrocyte mitochondrial dysfunction. After years of this process, the skin loses its moisture and support, appearing less lustrous and becoming more wrinkled and sagging, along with a multitude of other imperfections. The risk of skin cancer is also considerably increased.

AGEs, Mitochondria, and Skin

The activity of mitochondria, especially Complex II activity, is significantly impaired in older skin. In one recent study, Complex II activity was measured in twenty-seven donors ranging in age from 6 to 72 years. Samples were taken from a sun-protected area of skin to determine if there was a difference in Complex II activity with increasing age. It was found that Complex II activity per unit of mitochondria significantly declined with age in the cells derived from the deeper, rather than the upper, levels of skin, an observation not previously reported. The scientists found that the amount of enzyme protein decreased, observing this decrease only in those cells that had stopped proliferating (3).

As previously discussed, when glucose or aldehydes interact with skin proteins, they form glycated proteins. An additional problem is that the cross-linked proteins can interact with other oxidized molecules, resulting in larger and more harmful AGEs. These AGEs perpetuate damage to the skin collagen, elastin, and intracellular matrix and also induce fibroblast apoptosis.

AGEs are very potent stimulators of apoptosis. As a natural process, apoptosis eliminates damaged and useless cells through the process of autophagy. This creates space for new and healthy cells to develop. Aging itself causes increased rates of apoptosis because of critically shortened telomeres, elevated oxidized protein amyloid levels, and other cellular-suicide triggers. On the other hand, it's necessary to control and modulate the apoptotic rates of skin cells; otherwise, too many cells die and the skin structure becomes compromised.

Glycation is typically due to specific aldehydes, such as

malondialdehyde, 4-hydroxynonenyl, and methylglyoxal. These are byproducts of lipid peroxidation—the destruction of healthy lipids in cell and mitochondrial membranes, caused by free radicals, which was discussed in previous chapters. These cross-linking agents attack skin cell components, resulting in chronic skin microinflammation, which makes the skin appear red, rough, and itchy and also contributes to wrinkles and skin blemishes. The combined action of protein oxidation, glycation, and lipid peroxidation leads to chronic inflammation, resulting in visible signs of skin aging.

Other age-related changes to the skin include the following:

- the formation of age spots, which are oxidized protein aggregates similar to lipofuscin

- the increased likelihood of skin cancers, such as melanoma and squamous cell carcinoma

- thin and fragile skin that's easily bruised and has deformed capillaries and varicosities

Apart from the visible signs of aging, there are other, less visible changes. For example, age affects the sweat and sensory elements of the skin. Sweating becomes more problematic and less effective. As a consequence, efficient elimination of toxins from the skin declines, and toxins may thus accumulate in the body. Additionally, the sensory organs of the skin become less sensitive, and information about the external temperature as well as sensations of touch, pain, and stretching become less reliable.

Preventing and Treating Skin Aging

There's been considerable research into the effects of topically applied antioxidants to prevent or even reverse free-radical-induced skin injury. The skin has several protective mechanisms to protect it from externally produced free-radical damage. The epidermis is covered with a layer of sebum mixed with lipids to form a complex structure known as sink surface lipids (SSLs). These protect the underlying skin from externally generated free radicals attaching to the skin-cell membranes.

Research shows that the composition of SSL changes with age. Younger adults have a higher concentration of branched, monounsaturated fatty acids in the SSL compared with older people, meaning that the healthy concentration of protective fatty acids is compromised in old age. In addition, the concentration of antioxidants within the lipid barrier also changes with age. Antioxidants such as vitamin E, squalene, and coenzyme Q10 significantly decrease in older individuals (4).

Coenzyme Q10

Starting in a person's mid- to late 30s, the body becomes less able to synthesize sufficient coenzyme Q10 to meet its energy needs. This is worsened by bad eating habits, stress, infection, or certain drugs. Adequate levels of coenzyme Q10 can significantly suppress the expression of collagenase, an enzyme that destroys collagen.

Studies show that CoQ10, applied directly to the skin surface or taken internally, helps reduce the appearance of wrinkles and fine lines. It also promotes repair of fibroblasts. In addition, topically applied CoQ10 can penetrate the outer layers of the skin and exert antioxidant benefits deep in the dermis.

Recent work has shown that treatment of mitochondria cells deficient in coenzyme Q10 with exogenous CoQ10 can increase the output of the ETC nine-fold. Complexes II and III were shown to increase substantially over a control group of cells when added to neuronal cell cultures (5).

Vitamin E and Pycnogenol

Vitamin E and pycnogenol, a maritime pine bark extract, are potent antioxidants when applied topically to the skin or when administered orally. Both antioxidants improve the immune function of the skin and aid the skin's repair and regenerative processes. Pycnogenol has demonstrated anti-inflammatory, antioxidant, and capillary stimulation effects in in vitro and in vivo clinical studies. Because pycnogenol is a mixture of small molecules that are gallic acid equivalents, it can penetrate easily through the epidermis and reach the deeper dermis.

Ginkgo biloba extract is another natural mixture that improves skin microcirculation, helps prevent vascular imperfections, and acts as an antioxidant in a wide range of clinical studies.

Carnosine

Carnosine is a natural dipeptide produced endogenously. It consists of the amino acids alanine and histidine and works in three different ways to do the following:

- prevent lipid peroxidation in the skin-cell membranes
- reduce the risk of glycation-induced damage as a potent glycation inhibitor
- lower microinflammation of the skin by lowering inflammatory markers

Carnosine has been extensively investigated because of its wound-healing properties. The healing properties of carnosine are due to its antioxidant activity, its metal chelation ability, and its regulation of inflammation.

During wound healing, there's an overproduction of collagen

and skin fibroblasts. The balance between adequate and excessive collagen and fibroblast formation is quite delicate, and the natural wound-healing process in the skin often results in excessive scarring. Carnosine can modulate the production of factors that regulate the equilibrium of collagen and fibroblast production, thus reducing scar formation.

The same situation is true for age-related skin damage in which the body overproduces collagen and fibroblasts to replace skin fibroblasts lost through lysis or apoptosis. Carnosine is an essential nutrient in this respect.

The Eyes

Vision is a complex process in which photoreceptors in the eye gather visual information by absorbing light and sending electrical signals to other retinal neurons for initial processing and integration. The signals are then sent via the optic nerve to the brain, which ultimately processes the image and allows us to see.

The term "sending electrical energy" infers that energy is somehow generated and used up in the process. Transferring energy requires even more energy, which ultimately is generated by the mitochondria. The amount of information being received and processed by the eye requires a high concentration of mitochondria.

Photoreceptors, of which there are about 125 million in each eye, are specialized neurons that turn light into electrical signals. The two major types of photoreceptors are rods and cones. Rods are extremely sensitive to light and allow us to see in dim light, but they don't convey color. Rods constitute 95 percent of all photoreceptors in the human eye. Most of our vision, however, comes from cones, which work under varying light conditions and are responsible for acute detail and color vision.

The human eye contains three types of cones (red, green, and blue), each sensitive to a different range of colors. Because of their overlap in sensitivities, cones work in combination to convey information about all visible colors. You might be surprised to know that we can see thousands of colors using only three types of cones. Computer monitors use a similar process to generate a spectrum of colors.

The retina contains three organized layers of neurons. The rod and cone photoreceptors in the first layer send signals to the second layer—composed of interneurons—which relays signals to the third layer. The third layer consists of multilayered ganglion cells, which are specialized neurons near the inner surface of the retina. The axons of the ganglion cells form the optic nerve. Each neuron in the middle and third layer typically receives input from many cells in the first layer, and the number of inputs varies widely across the retina.

The central part of the human retina, where light is focused, is called the fovea, and it contains only red and green cones. The area around the fovea, known as the macula, is critical for reading and driving. Death of photoreceptors in the macula—that is, macular degeneration—is a leading cause of blindness among the elderly population in developed countries, including the United States.

Age-related macular degeneration (AMD) accounts for 54 percent of all blindness in Americans of European ancestry as well as 5 percent of all blindness globally (6). It affects 30 percent of people over the age of sixty-five (6, 7).

AMD is a progressive, degenerative disorder of the macula that results in central vision impairment (8). There are two subtypes of AMD: early and late AMD. Early AMD is characterized by moderate to severe, lipid-rich, subretinal pigment-epithelium deposits, called drusen, accompanied by pigment abnormalities. Individuals with late-stage AMD have vision loss due to a damaged macula in addition to the presence of drusen.

It's well established that photoreceptor cells and retinal ganglion cells are vulnerable to damage from high-energy photons. The reason for this vulnerability is they're rich in mitochondria, with the corresponding high-energy electron leakage. Retinal ganglion cells are unusual in that they have unmyelinated axons that are exposed to light as they traverse the retina. This situation is analogous to installing electrical wiring without a covering of insulation.

Unmyelinated axonal transmission demands large amounts of energy, reflecting the greater energy required to restore their unmyelinated axonal transmission potential. The combination of high-energy requirements and exposure to light makes unmyelinated axons vulnerable to auto-oxidation from free radicals generated internally and external photon damage.

To meet their high-energy demands, retinal ganglion cells have regular varicosities along their length that are packed with mitochondria. These cells are in a constantly energized state, and this creates a potent, free-radical-generating system, giving rise to oxidative stress that can result in glaucoma (9).

In the previous chapters on cardiovascular disease and neurodegeneration, we discussed the consequences of reduced blood flow and impaired toxin removal through the lymphatic system in areas of high mitochondrial density. These same conditions exist in the eye, where reduced blood flow to the optic nerve head restricts oxygen delivery to retinal ganglion cells. This is especially true in the periphery of the retina, resulting in a higher risk of glaucoma.

Exposure to light combined with low mitochondrial energy availability cause oxidative stress, making these cells vulnerable to excitotoxicity. Excitotoxicity is the process whereby a massive glutamate release in the central nervous system in response to trauma leads to the death of neurons. Mitochondria accumulate much of the calcium entering neurons via an overactivated N-methyl-D-aspartate receptor. This calcium accumulation plays a key role in the subsequent cell death caused by mitochondrial depolarization from the massive amount of glutamate present.

AGEs accumulate in the lens and the retina during the aging process. Crystallins, the major structural proteins of the lens that account for transparency, are highly susceptible to glycation and AGE cross-linking (9). Human lenses containing cataracts have higher levels of pentosidine and hydroimidazolone compared to clear lenses (10). Pentosidine is one of the AGEs formed by the glycation of proteins and is found in high concentrations in the lens and retina. Increased levels of pentosidine due to aging are a risk factor for cataract development (10).

Hydroimidazolone is an AGE formed by methylglyoxyl, a major cross-linking agent. RAGEs, an abbreviation for "receptors for AGEs," are also found in high levels in human cataracts (11). N-carboxymethyl-lysine, or CML, is another type of AGE found in cataracts and the vitreous humor of the eye. Abnormal levels of N-carboxymethyl-lysine is a predictor of cataract development, especially among diabetics (12). Both CML and RAGEs are present in the pathological lesions of age-related macular degeneration (11, 12). Older adults with age-related macular degeneration have higher levels of plasma CML and pentosidine compared with normal controls (13).

The Muscles

As we age, we may notice we're not as flexible as we used to be. Generally, you're told this is because you're not stretching enough. Stretching can help, but it's not the sole solution to relieving tight muscles.

My office building is next door to a high-rise retirement facility. Every morning when I drive to work, I pass groups of older people walking in the neighborhood. This always brings a smile to my face because two of the major tenets for creating healthy mitochondria are walking and breathing. However, I also note that most of these seniors exhibit a stiffness in their gait and many are walking with canes or walkers. This difference in gait becomes increasingly apparent as a person approaches old age.

Older adults have increased cross-linking of collagen and deposition of AGEs in skeletal muscle (14). In aging animals, including ourselves, cross-linking of collagen in muscle, tendons, and cartilage is associated with increased muscle stiffness, reduced muscle function, and accumulation of AGEs (14). AGEs may also play a role in sarcopenia (age-related muscle wasting) through upregulation of inflammation and endothelial dysfunction in the microcirculation of skeletal muscle. AGE-induced cross-linking of collagen is elevated in older adults and increases the stiffness of articular cartilage (16).

In a study of older, community-dwelling adults, elevated circulating N-carboxymethyl-lysine (CML) levels, one type of AGE that's also found in eye tissues, were independently associated with hip fracture risk, low grip strength (17), and slow walking speed (18).

Muscles contain some of the highest mitochondrial content of any tissue in the body because muscle cells must provide massive amounts of ATP for movement and exercise. Muscle is generally divided into three types—white muscle, red muscle, and mixed muscle. The terms "red" and "white" are derived from the way these muscles appear during surgery or autopsies, but they mainly refer to the mitochondrial content of the muscle itself.

Red muscles contain a large quantity of mitochondria, white muscles contain fewer mitochondria, and mixed muscles contain both red- and white-muscle fiber types.

Muscle cells often contain hundreds or even thousands of mitochondria to support the generation of large quantities of ATP during exercise. Strenuous aerobic exercise increases the mitochondrial numbers in heart and skeletal muscles.

The term mitochondrial biogenesis refers to the process of replicating mitochondria within a cell in order to increase ATP production in response to an increased demand for energy or from oxidative stress (19). Mitochondrial biogenesis results in an expansion of the network of mitochondria within a cell and an increase in the maximal amount of ATP that can be generated during intense exercise. More mitochondria result in more ATP production in peak exercise conditions (19).

Muscle Mitochondria: Use Them or Lose Them!

Chronic disuse of muscles, sedentary behavior, and aging contribute to a decline in mitochondrial content and function, leading to the production of free radicals and cell death. The muscle tissue of people with type 2 diabetes has also been extensively studied, revealing gross defects in mitochondrial number and function. Although the cause-and-effect relationship is unknown, muscle tissue from people with type 2 diabetes is often associated with reduced aerobic capacity, insulin resistance, and deficient mitochondrial biogenesis.

Studies have shown that defective mitochondrial biogenesis in the heart muscle can predispose individuals to cardiovascular complications, heart disease, and the risk of developing metabolic syndrome.

Fortunately, reversing the effects of aging, diabetes, and cardiovascular disease via increased mitochondrial biogenesis is as simple as exercising more. Studies have shown that in older individuals with existing metabolic disease, the resumption of an active lifestyle can significantly improve preexisting cellular damage and promote gains in muscle mass. Regular endurance exercise, by itself—independent of changes in diet—can normalize age-related mitochondrial dysfunction by stimulating mitochondrial biogenesis (20).

Exercise Is the Most Effective Way to Make New Mitochondria

Exercise is the most potent signal to increase production of mitochondria in muscle by stimulating the ability of the muscle to burn carbohydrates and fatty acids for ATP. When you exercise, muscle cells generate a low-energy signal known as AMP, and the accumulation of AMP over time signals an increase in ATP production. An increased ratio of AMP to ATP initiates a signal cascade within the muscle to produce more ATP, thereby protecting against an energy deficit. During periods of sustained muscle contraction, calcium is released from intracellular stores, resulting in a 300 to 10,000 percent increase in intracellular calcium. Increased calcium and AMP are powerful signals for increased mitochondrial biogenesis. Mitochondrial biogenesis takes place in the resting state immediately following exercise and is driven by temporary, oxidative, stress-induced free-radical production created by exercise (21).

During exercise, mitochondria consume larger amounts of oxygen, carbohydrates, and fatty acids—the fuels needed to power ATP production. The ability of muscles to overcompensate for exercise "stress" is exactly why frequent exercise results in increased strength, endurance, resistance to fatigue, and whole-body fitness. Lipid peroxidation rates increase enormously during strenuous aerobic exercise due to beta oxidation of fatty acids for energy and increased mitochondrial free-radical leakage that oxidizes cell and organelle membranes. Increased lipid peroxidation sends signals to the antioxidant response element (ARE), which stimulates increased antioxidant enzyme production and increased mitochondrial biogenesis to meet the new energy demands. PGC1-a gene expression activity is used to determine that mitochondrial biogenesis is taking place (22).

AGE Inhibitors for Skin, Muscles, and Eyes

The process of glycation cross-linkage was previously thought to be irreversible, and this was believed to be one typical example of the immutability of aging. There are now several commercially available inhibitors of cross-linking. Examples include carnosine, aminoguanidine, metformin, acarbose, benfotiamine, pyridoxamine, chebulic acid, and fisetin. Some of these (like acarbose and metformin) are already in use as antidiabetic drugs, but new research has found they have additional anti-cross-linking effects not previously known.

The first recognized compound to prevent and reverse already-formed AGEs is aminoguanidine. Although sold as a supplement, aminoguanidine has been found to have serious side effects at higher doses. Chebulic acid, found in *Terminalia chebula* extracts, removes 1,540 times more AGEs than aminoguanidine when compared in side-by-side tests (23).

Fisetin is a plant flavone that increases the level and activity of glyoxalase 1, the enzyme required for the removal of methylgyoxyl.

Benfotiamine and pyridoxamine prevent new AGE formation but don't remove existing AGEs from proteins.

Many cross-linking inhibitors are scavengers for reactive carbonyl intermediates. They're also copper chelators that minimize the risk of carbonyl formation, cross-linking, and consequent AGE-related damage.

The following is a list of some natural, well-tested antiglycation agents.

Carnosine

The dipeptide carnosine is a naturally occurring agent found in muscle and nervous tissue. Carnosine is one of the more promising cross-linking inhibitors. It has multiple actions and has been described as a pluripotent agent. Carnosine is one of the few cross-linking inhibitors that's not only active against protein-to-protein cross-linking but also against protein-to-DNA cross-linking (24).

Another important carnosine activity is "carnosinylation," a process by which carnosine attaches to a protein bearing a carbonyl group, thus blocking the carbonyl from attaching to another protein. This process is analogous to placing a piece of paper (carnosine) between two proteins bearing glue (carbonyls). In other words, carnosine reacts with carbonylated proteins to form carnosine-carbonyl-protein adducts. These adducts are then removed by DNA and protein repair enzymes to undergo proteolysis, with excretion of the resulting waste products in the urine and feces. Proteolysis is the breakdown of oxidized protein masses by the cell's proteolytic enzymes (25).

Carnosine has a direct antioxidant action, and it also has a sparing effect on other antioxidants such as glutathione. It's a strong chelator of copper, thereby reducing the copper-mediated free-radical damage that occurs during AGE activity (25).

At the clinical level, carnosine has been shown to reduce urinary end products of free-radical and glycosylation metabolism in humans (26). One of the most important developments regarding carnosine is its ability to ameliorate cataracts, glaucoma, and other age-related eye conditions. To treat these disorders, the form of carnosine used is N-acetyl-carnosine. The action of carnosine in the eye is perhaps related to its ability to stimulate proteolysis and thus dissolve protein aggregates in the lens (26).

Pyridoxamine

There are three different vitamers of vitamin B6: pyridoxal, pyridoxine (the form traditionally used in supplementation), and pyridoxamine. All of these are naturally occurring. Pyridoxamine (brand name Pyridox-Pro™) is derived from animal sources, whereas pyridoxine is found in plant sources. All three B6 analogs possess anti-cross-linking actions, but pyridoxamine is the strongest and most clinically significant. Therefore, it is a good candidate for treating the eyes, skin, and muscles to help minimize cross-linking coupled with minimizing carbohydrates and sugars that drive the cross-linking process.

Pyridoxamine reduces the formation of AGEs by 25 to 50 percent and ameliorates diabetes-related kidney dysfunction by improving albuminuria, plasma creatinine, and hyperlipidemia. It works by trapping reactive carbonyl groups and exhibits free-radical-scavenging properties.

Pyridoxamine doesn't affect blood glucose levels. It inhibits both methylglyoxal and glycoaldehydes, which are most active following lipid peroxidation. It forms methylglyoxal-pyridoxamine dimers that are inactive and easily eliminated. There have been reports of neurotoxicity from using very high doses of pyridoxine, but the use of pyridoxamine at normal human therapeutic levels are thought to be free of these side effects. Pyridoxamine needs to be phosphorylated before it can become active.

Summer Savory

In Chapter 10, which covers cancer, we discuss the work of the Morrés, who defined the class of ENOX proteins located on the cell surface. Specific to the cancer discussion, they identified a tumor-associated ENOX protein and specific nutraceuticals to close the protein doors located on its cell surface.

The Morrés also identified a protein called ENOX3, the age-related ENOX (abbreviated arNOX) that's similar in its fundamental structure to ENOX but originates from genes on different chromosomes (27). ENOX3 begins to appear at around age thirty and then increases after age sixty or seventy. The amount of ENOX3 in blood or saliva correlates quite strongly with chronological age. In other words, the more ENOX3 one has, the older one looks.

The arNOX proteins propagate ROS to surrounding cells and circulating lipoprotein particles, which occurs during skin aging and atherogenesis. arNOX is widely distributed among aging systems, including late-passage cultured cells and plants. By oxidizing proteins in the supporting matrices that are important to skin health, elevated arNOX activity has been demonstrated to be a major cause of skin aging (28).

In the laboratory, the Morrés discovered that certain, readily available nontoxic substances strongly inhibit the destructive effects of ENOX3, suppressing one of the mainsprings of the aging process. Coenzyme Q10 and salicin used in skin care and oral supplements inhibit ENOX3 activity by reducing lipoprotein oxidation, which has been shown to lead to coronary heart disease.

The Morrés determined that subjects with high arNOX activity had skin characteristics that made them appear, on average, seven years older than their chronological age. Subjects with low arNOX activity at the same age had skin that appeared to be seven years younger than their actual age (29).

Another surprising fact the Morrés discovered is that certain herbs are protective against ENOX3. One herbal decoction is the famous French mixture, herbs de Provence. This a traditional blend of dried herbs found in the Provence region of southeast France. The Morrés discovered that the most effective and beneficial of these herbs is summer savory, *Satureja hortensis*. These Provencal herbs have very active biological effects in humans, even at the minute levels used in French cooking (30).

The Morrés found that, by incorporating the herbal preparations as sustained-release formulations, all-day protection could be attained with just two, 400 mg capsules per day in divided doses. This would maintain arNOX levels in an aging population at nearly those of a thirty-year-old person (30).

Summary

Aging of the skin, eyes, and muscles occurs through cross-linking by aldehydes, protein oxidation, and lipid peroxidation, which are the three stochastic (random) mechanisms that modify the body's proteins. All three processes produce aldehydes, which cross-link body proteins in an unscheduled, unnatural manner.

Cross-linking is somewhat analogous to hand-cuffing every third worker in a factory. Production slows down and productivity declines, which is what we observe in aging organisms. Muscle stiffness develops and arteries become less flexible, leading to hypertension and poor flow-mediated dilation. Skin becomes tougher, leathery, and wrinkled.

Aldehydes produced during glycation include methylglyoxyl, pentosidine, and N-carboxymethyl lysine (CML), which are found in cataractous tissues and the vitreous humor of the eye but are also found in all other aged tissues to some extent.

Beta amyloid, found most notably in the brain and the driving force in Alzheimer's disease, is present in all other aging, post-mitotic cell types, too.

Lipid peroxidation is the free-radical-induced destruction of cell, organelle, and mitochondrial membranes.

Proteins are modified by protein oxidation to form carbonyls, which serve as an attachment platform for building AGEs.

The three processes of glycation, lipid peroxidation, and protein oxidation work together to accelerate the aging process in a synergistic manner in which they cause mitochondrial dysfunction, mitochondrial DNA mutations, genomic instability, and nuclear DNA mutations.

These modifications of body proteins result in cancer, Alzheimer's disease, and other degenerative diseases of aging as well as skin wrinkling, muscular stiffness, cataracts, and other age-related, degenerative eye conditions.

In the next chapter, we'll discuss cancer, another degenerative, age-related disease influenced by mitochondrial dysfunctions.

References:

1. Ergin, V., Hariry, R.E., and Karasu, C. "Carbonyl Stress in Aging Process: Role of Vitamins and Phytochemicals as Redox Regulators." *Aging and Disease* 4, no. 5 (2013): 276–94.
2. Ansari, N.A., Moinuddin, Mir, A.R.; Habib, S., Alam, K., Ali, A., and Khan, R.H. "Role of Early Glycation Amadori Products of Lysine-Rich Proteins in the Production of Autoantibodies in Diabetes Type 2 Patients." *Cell Biochemistry and Biophysics* 70, no. 2 (2014): 857–65
3. Bowman, A. and Birch-Machin, M.A. "Age-Dependent Decrease of Mitochondrial Complex Activity in Human Skin Fibroblasts." *Journal of Investigative Dermatology* 136, no. 5 (May 1, 2016): 912–9.
4. Passi, S., De Pita, O., Puddy, P., and Litarru, G.P. "Lipophilic Antioxidants in Human Sebum and Aging." *Free Radical Research* 36, no. 4 (April 2002): 471–7.
5. Duberley, K.E., Heales, S.J.R., Abramov, A.Y., Chalasani, A., et al. "Effect of Coenzyme Q10 Supplementation on Mitochondrial Electron Transport Chain Activity and Mitochondrial Oxidative Stress in Coenzyme Q10 Deficient Human Neuronal Cells." *The International Journal of Biochemistry and Cell Biology* 50 (May 2014): 60–3.
6. Pascolini, D. and Mariotti, S.P. "Global Estimates of Visual Impairment." *British Journal of Ophthalmology* 96, no. 5 (2012): 614-8.
7. Congdon, N., O'Colmain, B., Klaver, C.C., et al. "Causes and Prevalence of Visual Impairment among Adults in the United States." *Archives of Ophthalmology* 122, no. 4 (2004): 477–85.
8. Restrepo, N.A., Mitchell, S.L., Goodloe, R.J., et al. "Mitochondrial Variation and the Risk of Age-Related Macular Degeneration across Diverse Populations."

Pacific Symposium on Biocomputing (2015): 243–54

9. Osborne, N.N., Lascaratos, G., Bron, A.J., Chidlow, G., and Wood, J.P.M. "A Hypothesis to Suggest That Light Is a Risk Factor in Glaucoma and the Mitochondrial Optic Neuropathies." *British Journal of Ophthalmology* 90 (2006): 237–41.

10. Kumar, P.A., Kumar, M.S., and Reddy, G.B. "Effect of Glycation on Alpha-Crystallin Structure and Chaperone-like Function." *Biochemical Journal* 408, no. 2 (December 1, 2007): 251–8.

11. Franke, S., Dawczynski, J., Strobel, J., Niwa, T., Stahl, P., and Stein, G. "Increased Levels of Advanced Glycation End Products in Human Cataractous Lenses." *Journal of Cataract and Refractive Surgery* 29, no. 5 (May 2003): 998–1004.

12. Hammes, H.P., Hoerauf, H., Alt, A., Schleicher, E., Clausen, J.T., Bretzel, R.G., and Laqua, H. "N-(epsilon)(carboxymethyl)lysin and the AGE Receptor RAGE Colocalize in Age-Related Macular Degeneration." *Investigative Ophthalmology and Visual Science* 40, no. 8 (July 1999): 1855–9.

13. Ni, J., Yuan, X., Gu, J., Yue, X., Gu, X., Nagaraj, R.H., Crabb, J.W., and The Clinical Genomic and Proteomic AMD Study Group. "Plasma Protein Pentosidine and Carboxymethyllysine, Biomarkers for Age-Related Macular Degeneration." *Molecular and Cellular Proteomics* 8, no. 8 (August 2009): 1921–33.

14. Haus, J.M., Carrithers, J.A., Trappe, S.W., and Trappe, T.A. "Collagen, Cross-Linking, and Advanced Glycation End Products in Aging Human Skeletal Muscle." *Journal of Applied Physiology* 103, no. 6 (December 2007): 2068–76.

15. Wood, L.K., Kayupov, E., Gumucio, J.P., Mendias, C.L., Claflin, D.R., and Brooks, S.V. "Intrinsic Stiffness of Extracellulular Matrix Increases with Age in Skeletal Muscle of Mice." *Journal of Applied Physiology* 117,

no. 4 (August 15, 2014): 363–9.

16. Couppé, C., Hansen, P., Kongsgaard, M., Kovanen, V., Suetta, C., Aagaard, P., Kjaer, M., and Magnusson, S.P. "Mechanical Properties and Collagen Cross-Linking of the Patellar Tendon in Old and Young Men." *Journal of Applied Physiology* 107, no. 3 (September 2009): 880–6.

17. Dalal, M., Ferrucci, L., Sun, K., Beck, J., Fried, L.P., and Semba, R.D. "Elevated Serum Advanced Glycation End Products and Poor Grip Strength in Older Community-Dwelling Women." *Journal of Gerontology: Series A, Biological Science and Medical Sciences* 64, no. 1 (January 2009): 132–7.

18. Semba, R.D., Bandinelli, S., Sun, K., Guralnik, J.M., and Ferrucci, L. "Relationship of an Advanced Glycation End Product, Plasma Carboxymethyl-Lysine, with Slow Walking Speed in Older Adults: The InCHIANTI Study." *European Journal of Applied Physiology* 108, no. 1 (January 2010): 191–5.

19. Zhang, Y. and Xu, H. "Translational Regulation of Mitochondrial Biogenesis." *Biochemical Society Transactions* 44, no. 6 (December 15, 2016): 1717–24. Review.

20. Greggio, C., Jha, P., Kulkarni, S.S., Lagarrigue, S., et al. "Enhanced Respiratory Chain Super Complex Formation in Response to Exercise in Human Skeletal Muscle." *Cell Metabolism* (November 28, 2016). pii: S1550-4131(16)30582-4.

21. Silvennoinen, M., Ahtiainen, J.P., Hulmi, J.J., Pekkala, S., et al. "PGC-1 Isoforms and Their Target Genes Are Expressed Differently in Human Skeletal Muscle Following Resistance and Endurance Exercise." *Physiological Reports* 3, no. 10 (October 2015). pii: e12563.

22. Wang, L., Mascher, H., Psilander, N., Blomstrand, E., and Sahlin, K. "Resistance Exercise Enhances the Molecular Signaling of Mitochonbdrial Biogenesis

Induced by Endurance Exercise in Human Skeletal Muscle." *Journal of Applied Physiology* 111, no. 5 (November 2011): 1335–44.

23. Lee, J.Y., Oh, J.G., Kim, J.S., and Lee, K.W. "Effects of Chebulic Acid on Advanced Glycation End Products-Induced Collagen Cross-Links." *Biological and Pharmaceutical Bulletin* 37, no. 7 (2014): 1162–7.
24. Pepper, E.D., Farrell, M.J., Nord, G., and Finkel, S.E. "Antiglycation Effects of Carnosine and Other Compounds on the Long-Term Survival of Escherichia Coli". *Applied and Environmental Microbiology* 76, no. 24 (December 2010): 7925–30.
25. Boldyrev, A.A., Aldini, G., and Derave, W. "Physiology and Pathophysiology of Carnosine." *Physiological Reviews* 93, no. 4 (October 2013): 1803–45.
26. Liao, J.H., Lin, I.L., Huang, K.F., Kuo, P.T., Wu, S.H., and Wu, T.H. "Carnosine Ameliorates Lens Protein Turbidity Formations by Inhibiting Calpain Proteolysis and Ultraviolet C-induced Degradation." *Journal of Agricultural Food Chemistry* 62, no. 25 (2014): 5932–8.
27. Morré, D.J. and Morré, D.M. *ECTO-NOX Proteins: Growth, Cancer, and Aging*, (New York: Springer, 2013).
28. Morré, D.J. and Morré, D.M. *ECTO-NOX Proteins: Growth, Cancer, and Aging,* (New York: Springer, 2013).
29. Morré, D.J. and Morré, D.M. *ECTO-NOX Proteins: Growth, Cancer, and Aging,* (New York: Springer, 2013).
30. Morré, D.J. and Morré, D.M. *ECTO-NOX Proteins: Growth, Cancer, and Aging*, (New York: Springer, 2013).

Chapter 10

The Role of Mitochondrial Dysfunction in the Development of Cancer

"All normal cells have an absolute requirement for oxygen, but cancer cells can live without oxygen—a rule without exception."
--The Metabolism of Tumours, Otto H. Warburg, 1930

The hallmark of cancer is unregulated cell growth, and a fundamental question is how this is accomplished. Countless billions of dollars have been spent over the last fifty years in efforts to identify the cellular and genetic pathways involved in this aggressive growth of abnormal cells.

In this chapter, we'll discuss the role that defective mitochondria have in the development of cancer, and we'll see that key proteins on the surface of cancer cells are different than those on normal cells and play a critical role in cancer cell metabolism.

All human cells, including normal and transformed cells, require energy for cellular metabolism, growth, repair, cell division, and, eventually, apoptosis. In normal cells, about 95 percent of that energy comes from our mitochondria and the balance from cellular glycolysis. Cancer cells are just the opposite—they depend primarily on anaerobic glycolysis, a cytosolic fermentation process that occurs outside the mitochondria—for their energy needs.

Mitochondria are thought of as the cell's power plants, and ATP is the energy currency they generate for the body to use. In normal cells, this energy is generated through oxidative respiration via two pathways: fermentation (glycolysis) and respiration (cellular respiration requiring oxygen). Glycolysis takes place in the interior of the cell, the cytosol, which doesn't require any participation in the Krebs cycle or the mitochondrial respiratory chain. It's an anaerobic process that begins with one molecule of glucose and, through a series of ten steps, transforms that molecule into two molecules of pyruvate.

Once pyruvate is generated, the cell has a decision to make. It can take pyruvate into the mitochondria, where it will begin the respiratory energy cycle—the highly efficient process that employs oxygen to generate twenty-three ATP molecules. Alternately, the cell can ferment the pyruvate, an inefficient method of energy production that produces only two molecules of ATP and generates a byproduct: lactate or lactic acid.

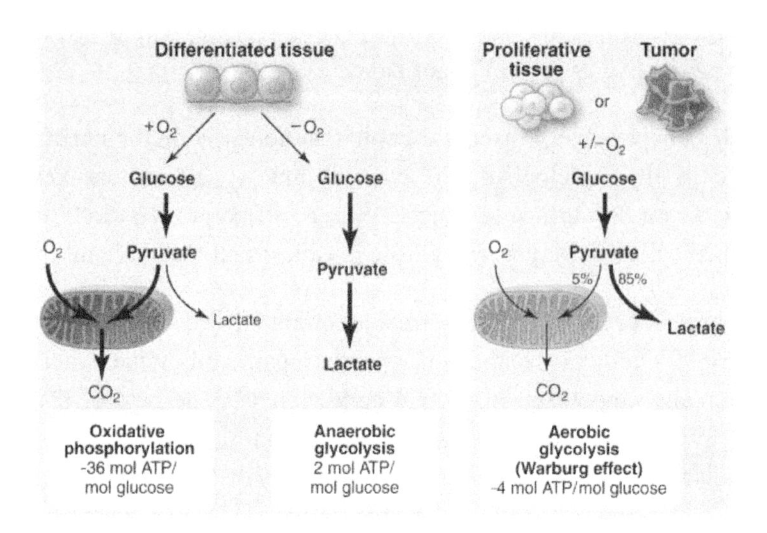

Figure 7. Schematic representation of the differences between oxidative phosphorylation, anaerobic glycolysis, and aerobic glycolysis (Warburg Effect).
Credit:
https://www.ncbi.nlm.nih.gov/pmc/articles/PMC2849637/

In the early 1900s, Otto Warburg became the first person to identify biochemical alterations of cancer cells characterized by a shift in glucose metabolism from oxidative phosphorylation to glycolysis. Warburg's observation that cancer cells use glycolysis to meet their energy requirements was largely ignored by the mainstream scientific community. Later, this observation was named the Warburg effect.

Warburg also observed that cancer cells tend to convert most glucose to lactate regardless of whether oxygen is present (aerobic glycolysis). This property is shared by normal proliferative tissues. Mitochondria remain functional, and some oxidative phosphorylation continues in both cancer cells and normal proliferating cells. Nevertheless, aerobic glycolysis is less efficient than oxidative phosphorylation for generating ATP. In proliferating cells, approximately 10 percent of the glucose is diverted into biosynthetic pathways upstream of pyruvate production.

Tumor cells that are actively proliferating have been observed to undergo a switch from oxidative to glycolytic metabolism. Tumors characteristically metabolize most of the glucose they receive through glycolysis even in the presence of an adequate oxygen supply. However, tumors have a substantial reserve capacity to produce ATP by oxidative phosphorylation when glycolysis is suppressed. A high rate of glycolysis is required to support cancer cell growth to compensate for defects in mitochondrial function (1). Tumor cells tend to use a lot more glucose than normal cells because glycolysis is a much less efficient method of converting glucose into ATP than oxidative phosphorylation. This shift is a growth advantage to cancer cells because glycolysis doesn't require oxygen. Cancerous tumors frequently outgrow their blood supply; they often live and grow in tissue spaces where there isn't much oxygen.

The preference of cancer cells for glucose has been known for a long time and is the basis for positron emission tomography (PET) scanning. PET scanning uses a radiolabeled derivative of glucose as a tracer because tumor cells take up radiolabeled glucose much more aggressively than normal cells do. These high-glucose-uptake areas show up as brightly colored blobs on PET-CT (computed tomography) scans, indicating the location of tumor masses (2).

There's intense interest in understanding what role the Warburg effect plays in cancer initiation and growth, and therapeutics are being developed in the hope of reversing metabolic conditions required by cancer cells. Warburg originally hypothesized that cancer cells develop a defect in mitochondria that leads to impaired aerobic respiration and a subsequent reliance on glycolytic metabolism. However, later work showed that mitochondrial function isn't impaired in most cancer cells, suggesting an alternative explanation for aerobic glycolysis in cancer cells.

To this end, the potential role of dietary supplements and tight glucose control as adjuncts to cancer treatment is an active field of investigation (3). Sulforaphane—a naturally occurring isothiocyanate found in cruciferous vegetables, such as broccoli, brussels sprouts, and cabbage—has received a great deal of attention because of its ability to inhibit cell proliferation and induce apoptosis in the mitochondria of cancer cells. A body of evidence shows that this phytochemical inhibits growth and induces apoptosis in many different types of cancer cells in humans (4, 5, 6).

In order to kick-start the mechanism to induce apoptotic mitochondrial death in cancer cells, sulforaphane must release the protein cytochrome c. The intermembrane space in mitochondria contains many proapoptotic proteins, including cytochrome c. Release of cytochrome c in the intermembrane space drives mitochondrial apoptosis. Recent research suggests that sulforaphane-induced apoptosis in cancer cells occurs via the mitochondrial pathway. The results of numerous experimental models (7, 8, 9) have demonstrated the ability of sulforaphane to cause inhibition of cell growth and induction of apoptosis in cancer cells. Based on these observations, it's reasonable to conclude that sulforaphane could be used as an adjunct in cancer treatment for inducing apoptosis.

Metabolic Inflexibility Resulting from a Poor Diet

The ability of cells to switch between the two types of substrate utilization is essential to survival. This switch occurs when fat stores are used by muscles during periods of starvation. The importance of fuel selection is underscored by studies indicating that "metabolic inflexibility," or impaired capacity to switch between nutrient utilization has a pathogenic role in the insulin resistance commonly seen in type 2 diabetes, obesity, and degenerative diseases associated with aging.

A lifetime of poor dietary choices predisposes an individual to substrate utilization inflexibility and this, coupled with a lack of exercise, drives age-associated diseases, including cancer (16). One option to reintroduce metabolic flexibility to the cells is to adopt a ketongenic or calorie-restricted diet.

The Benefits of Ketogenic and Calorie-Restricted Diets

If glucose is the primary fuel of cancer, it would seem reasonable to investigate ways to minimize the intake of glucose derived from sugars and carbohydrates. Glucose may serve as a primary energy source for cancer, but cancer cells are resilient to glucose deprivation. Cancer cells back up their energy resources by two means: the conversion of lactate to ATP and the use of the mitochondrial oxidative phosphorylation (OXIPHOS) pathway.

Thomas Seyfried, PhD, the author of the book *Cancer as a Metabolic Disease*, follows Warburg's original line of thinking that cancer is a metabolic disorder. Seyfried notes that, across the board, cancer cells do indeed exhibit extensive mitochondrial damage (10). Cells with damaged mitochondria switch over to glycolysis when they're unable to generate enough energy for survival due to impaired oxidative respiration. Once this switch is made, the cells begin to exhibit the hallmark features of cancer: uncontrolled proliferation, genomic instability (the increased probability that DNA mutation will occur), evasion of cell death, and so forth (10).

According to Seyfried, the damaged mitochondria instigate the energy switch to glycolysis. This leads to further mutations to the mitochondrial DNA (mtDNA), which are thought to precipitate and drive the cancer. In other words, metabolic forces, not genetics, are the primary driver of cancer.

In experiments involving tumors in rodents, Seyfried observed that the faster-growing tumors had lower overall numbers of mitochondria and resorted to increased glucose fermentation for energy. Additional evidence showed that the mitochondria in the cancer cells that grew the fastest were rife with a spectrum of structural abnormalities. The cells were smaller, less robust, cup- or dumbbell-shaped, missing important internal membranes, and had numerous abnormalities in their protein and lipid content. In biology, structure equals function. If the mitochondria of cancer cells are structurally altered, one can be sure their energy-production function will be reduced (11).

Seyfried notes in his work with rodent tumors that simple calorie restriction caused the tumors to shrink. Calorie restriction not only limits caloric intake but also lowers glucose levels, and it has been shown to have beneficial results in cancer studies (12).

This approach of limiting the nutrients needed for glucose production forces cancer cells to compete with healthy cells for the fuel they need. Additionally, with the anaerobic pathway of glucose limited, the cancer cells must use the mitochondrial OXIPHOS energy pathway instead (13).

It would make sense that cancer cells that have defective mitochondria would also have limited cell growth and cell division when they're deprived of glucose. The expected result would be a slowing of tumor growth. Unfortunately, there's no clinical data to support this approach to cancer treatment.

Surprisingly, long-term calorie restriction isn't recommended as a cancer treatment because the potential risk of cachexia (undesired weight loss) would force clinicians to reduce the dose of chemotherapy used, thereby decreasing the efficiency of the treatment.

This is one of the main reasons why the current recommendation of the American Cancer Society is to increase food intake for cancer patients receiving chemotherapy (14).

Recently, scientists conducted a small, Phase I clinical trial in which patients undergoing chemotherapy fasted for seventy-two hours. The results were promising; the patients' hematopoietic stem cells were protected from the negative side effects of the chemotherapy when compared with the nonfasting control group

A dietary protocol Seyfried developed showed promise in its ability to slow the growth of cancer and work synergistically with existing therapies. This protocol would also mitigate side effects of current cancer treatments. Seyfried's protocol recommended cutting out vegetables, bread, pasta, grains, and sugar while increasing the consumption of foods high in fat, such as nuts, cream, and butter. He modified the diet slightly, keeping overall calories restricted but eliminating carbohydrates in favor of fats—a modification that may put even more metabolic pressure on cancer cells (10).

With no carbohydrates available, the body is forced out of its normal method of metabolic energy generation and automatically manufactures molecules called ketone bodies to take the place of glucose as a source of circulating fuel. Ketone bodies are produced by the liver from fatty acids and are used as an energy source when glucose isn't readily available (10).

Background of the Ketogenic Diet

The ketogenic diet is high in fat and adequate in protein but low in carbohydrate. It's used to treat epilepsy in children (15). The diet designed for the treatment of epilepsy forces the body to burn fats rather than carbohydrates. Normally, the carbohydrates contained in food are converted into glucose, which is then transported around the body and is particularly important in fueling brain function. However, if there's very little carbohydrate in the diet, the liver converts fat into fatty acids and ketone bodies. The ketone bodies pass into the brain and replace glucose as an energy source. For children with epilepsy, an elevated level of ketone bodies in the blood—a state known as ketosis—leads to a reduction in the frequency of seizures (15).

The ketogenic diet contains a 4:1 ratio of fat to combined protein and carbohydrate. This is achieved by excluding foods such as starchy fruits, vegetables, simple sugars, and refined carbohydrates.

The Calorie-Restricted Diet

Another dietary alteration that's commonly chosen to return metabolic flexibility and reverse the effects of chronic disease is calorie restriction. Calorie restriction slows the aging process and increases the maximum lifespan of various short-lived animal species. Alternatively, high-fat feeding is associated with insulin resistance and may lead to an oversupply of fatty acids in the mitochondria. In humans, high-fat feeding results in downregulation of several genes associated with oxidative phosphorylation and mitochondrial biogenesis.

Calorie restriction has many beneficial physiological benefits in nonhuman primates. It also has positive physiological benefits in obese humans aside from simple weight loss. Very little is known, however, regarding the effects of long-term calorie restriction paired with good nutrition in humans who are in a normal weight range. The range of calorie restriction recommended is modest—reducing caloric intake by 10 to 20 percent of normal levels.

A randomized, controlled trial was conducted to determine the feasibility of prolonged calorie restriction in nonobese adults. This was compared the effects of a group undergoing calorie restriction with exercise-induced weight loss to produce a similar energy deficit in a healthy-lifestyle control group. The study, entitled CALERIE, was "the first clinical trial of calorie restriction in non-obese humans, and the present study is the first that directly compares the effects of a long-term, diet-induced energy deficit with a comparable exercise-induced energy deficit on whole-body and abdominal adiposity. The results support the feasibility of long-term, modest CR (calorie restriction), and demonstrate that one year of CR reduces whole-body adiposity and abdominal adipose tissue significantly and comparably to exercise. Novel features of this study, as compared to many previous studies of energy restriction, are the relatively long duration of the intervention and the focus on non-obese individuals" (17).

In a well-fed, sedentary person, mitochondria have plenty of fuel, but the cells don't use the ATP that's being generated. ATP levels remain high with little turnover. With this low demand for ATP, the electron transport chains (ETCs) become backed up with excessive electrons. Since there's still an abundance of oxygen and highly reactive electrons, there's a high rate of free-radical leakage. This burst of free radicals exceeds the built-in antioxidant defense system and oxidizes the lipids in the mitochondrial membranes (18). Such damage to mitochondrial membranes causes serious wear on the mitochondria and contributes to aging of the cells and the entire body, as discussed in Chapter 3.

This damage results in the release of cytochrome c from the inner mitochondrial membrane into the intermembrane space. When this happens, electron flow down the ETC completely stops (18). Now the upstream sections of the ETC are brimming with electrons, and these continue to leak and form more free radicals. Once this stress crosses a threshold, pores in the outer mitochondrial membrane open and initiate the first steps of apoptosis (18).

Membrane Transporters

Cells are complex structures that maintain tight homeostatic controls over their interior environment. Examples of this control are tightly regulated cellular pH, the ratio of NAD +/ NADH, and ROS. Inability to regulate the cellular environment drives apoptosis both in the mitochondria and the cell itself.

Membranes provide the cell with the essential ability to delineate the unregulated external environment from the specific, homeostatically controlled interior of the cell. Within the cell, compartments can be further subdivided and assigned specialized functions. This separation is essential for generating and utilizing electrical potential via regulated ion current, protecting precious replicative information from mutagenic insults, enforcing localization of molecules, and conversion of high-energy electrons into high-energy phosphates using proton flow.

The benefits of separable intracellular compartments are only truly achieved when the transport of molecules across membranes is regulated. This regulation occurs by a variety of mechanisms and the term "membrane transporter" is commonly used to describe most proteins that facilitate movement across a membrane (19).

The possibility that some cancer cells rely on membrane transporters for survival may offer an opportunity to target protein membrane transporters with nutraceuticals to disrupt the growth of cancer cells. ECTO-NOX (ecto-nicotinamide dinucleotide oxidase disulfide-thiol exchanger) proteins are just such a group of membrane transporters. ECTO-NOX proteins, also called NOX proteins, are cell-surface-associated and growth-related NADH oxidase enzymes that also transfer electrons to oxygen. This electron transport by cell-surface NOX proteins is called PMET (plasma membrane electron transport) (20, 21). Over the past twenty-two years, NOX family members have been identified as important contributors to many signaling pathways. They're found on the surface of cell membranes and immune cells. NOX proteins have been shown to regulate many fundamental physiological processes, including cell growth, differentiation, and apoptosis (20).

The similarity between the NAD/NADH inside the mitochondria and NOX cell-surface proteins is this: NADH is the reduced form of NAD; NADH has one more electron than NAD. NADH oxidation occurs inside the mitochondrial electron transport chain for cellular energy production in which three more electrons are added to it to make ATP and water. Anaerobic glycolysis allows glucose metabolism to progress by cycling NADH back to NAD+ through the transfer of electrons. NADH is converted to NAD+ to drive the electron transport chain in the mitochondria, which ultimately leads to the production of ATP in the cell.

On the external cell membrane, however, NADH plays a very different role—it oxidizes oxygen by adding only one electron to it to generate a superoxide radical, which isn't used to generate energy. In PMET, the cell must maintain a stable ratio of NADH to NAD for homeostasis. Normally, under general conditions, the cell is able to maintain a homeostatic balance of the NADH/NAD ratio.

Despite their beneficial roles, NOX cell-surface proteins also have a dark side. The superoxide radicals they generate steal electrons from adjacent cell membranes, which are made up of easily oxidized fatty acids. The superoxide radical removes a hydrogen from a fatty acid, setting in motion the free-radical cascade known as propagation. The fatty acid attacked now has an unpaired electron and bonds with oxygen to create a peroxyl radical, which steals an electron from the next adjacent fatty acid until eight to ten molecules of the cell membrane are destroyed (21).

This destruction of cell membranes is called lipid peroxidation; it was the first major aspect of the aging process that was delineated from 1958 to 1965 (21, 22). NOX-protein superoxide generation can spread free radicals throughout the body. These high-energy radicals can oxidize LDL cholesterol, which then leads to foam-cell formation and atherosclerosis (23, 24). Superoxide generation can trigger the formation of tau proteins and beta-amyloid proteins in brain cells, eventually causing Alzheimer's disease (25). Superoxide radicals can also destroy dopaminergic neurons in the brain, which can result in Parkinson's disease (25).

NOX Cancer Cell Variants

A cancer-cell variant of NOX proteins, discovered in 1992, is ENOX, or tNOX (tumor-associated NOX) in the literature (26). Numerous studies in the 1990s correlated NOX protein NADH oxidase activity with controlling cell growth. This cell-growth control mechanism in normal cells requires growth factors and hormones to activate it. The list of growth factors and hormones that upregulates NOX activity includes insulin, glucagon, EGF (epidermal growth factor), vasopressin, and lactoferrin (23, 27).

Plasma membranes of normal cells have baseline levels of NOX NADH oxidase, and this enzyme doubles its activity as a result of the action of these hormones and growth factors (27, 28). In contrast, cancer cell membrane ENOX or tNOX proteins already have upregulated levels of NADH oxidase and aren't further stimulated by the addition of growth factors. This lack of stimulation of NADH oxidase ENOX proteins on cell membranes is unique to cancer cells (29). This means that the NADH oxidase of cancer cell membranes is no longer correctly coupled to growth factors or hormones. This biochemical defect parallels the loss of growth control characteristics of fully transformed cancer cells (29, 30). How this change in cell-membrane NADH oxidase activity occurs and what's responsible for it remains to be determined by future research.

The takeaways here are that the NADH oxidase in the membranes of cancer cells is automatically upregulated to twice its normal levels and doesn't require growth-factor stimulation to achieve this. Cancer cell membranes are more fragmented and have an NADH oxidoreductase that also may be important in the regulation of cell growth. Cancer cell membrane NADH is somehow coupled to active growth factor receptors, but how this occurs is still under investigation (26, 27, 28, 29, 30, 31).

Cancer cell tNOX proteins are responsive to chemotherapeutic drugs. However, a variant of tNOX proteins called CNOX is resistant to chemotherapeutic drugs (29). There's also an age-related NADH oxidase called arNOX found on the membranes of aged cells that have some characteristics in common with cancer cell tNOX proteins. ArNox was discussed in depth in the previous chapter when we discussed the effects of mitochondrial dysfunction on skin, eyes, and muscles (32).

Tumor-associated tNOX proteins have a variant called ENOX2 that has an external receptor site that doesn't require chemotherapeutic drugs to pass through the cell membrane and enter the cell to be effective—they can simply bind to this outside receptor to exert their cancer-killing effects through the receptor (33). This makes ENOX2 an attractive target for the development of chemotherapy drugs because it has specific sensitivity to inhibition by quinone-site receptor drugs, which include nonsteroidal anti-inflammatory drugs (NSAIDs)—ibuprofen, aspirin, sodium naproxen, suramin, cis-platinum, sulforaphane, the green tea flavonoid EGCG, vitamin D3, the flavonoid phenoxodiol, doxorubicin, capsaicin, and the antitumor sulfonylureas (34, 35, 36). When cancer cells were exposed to the natural compounds EGCG and phenoxodiol at therapeutic levels of one hundred micromoles and ten micromoles, respectively, cell enlargement ceased and created a population of small cells that were unable to divide further and underwent apoptosis (35).

ENOX2 inhibitors block cell enlargement, which also block cell proliferation because, when a cell divides, it must reach a certain minimum size before it can divide again.

Inhibiting ENOX2 leads to growth inhibition and, after a few days, allows the apoptosis process of cellular suicide to take place (38). The growth-arresting properties of chemotherapy drugs on cancer cells are due to the blockage of ENOX2-catalyzed protein disulfide-thiol interchange, which is mandatory for cell proliferation (38).

The Glycolysis Shift and ENOX Proteins

The plasma-membrane electron transport that takes place through already highly upregulated tNOX cancer cell-surface proteins is key to understanding how cancer cells thrive when they shift to glycolytic energy production. One must keep in mind that glycolysis is far less efficient than energy production through oxidative phosphorylation. Glycolysis produces only two ATP molecules per glucose molecule, while oxidative phosphorylation produces more than thirty ATP molecules (1, 40).

Not all tumor cells shift to glycolytic energy production, but the most glycolytic tumors are the most aggressive tumors. (33, 41).

Some cancer cells probably rely almost exclusively on PMET for survival, and this may be an opportunity to use PMET as a target for anticancer therapeutics. PMET is pertinent to cell survival because it maintains various homeostatic relationships—such as the ratio of NAD+/NADH—that are essential for a cell to stay alive. If the electron transport chain is blocked through PMET by inhibition of tNOX, the NAD+/NADH levels will be offset and NADH levels will rise. NAD+ is necessary to serve as a redox agent for the electron transport chain, so, without adequate levels, the mitochondria won't sustain proper energy production.

Many aggressive and invasive cancers rely on glycolysis for their energy requirements, and enhanced glycolytic rates involve increased PMET activity to reduce stress caused by a buildup of NADH inside the cancer cells. However, the generation of free radicals by PMET on the surface of these cancer cells exports superoxide free radicals all over the body, causing damage to neighboring cells in a cascading effect (42). Mitochondrial DNA mutations allow these cells to survive by switching to glycolysis, with oxidative phosphorylation no longer an option (42).

Achieving PMET inhibition by reducing ENOX2 activity using drugs and natural compounds is an ongoing research target. Researchers are trying to identify the precise mechanisms of cancer cell-specific growth inhibition. Among the natural PMET inhibitors is a class of compounds that includes various capsaicins found in fruit peppers of the *Capsicum* species that acts on vanilloid receptors. Capsaicin is a more widely studied vanilloid that acts as a PMET inhibitor to induce apoptosis in cancer cells by inhibiting plasma membrane tNOX2 (43).

It's noteworthy that ENOX2 inhibition is achieved at low nanomolar concentrations (5 to 10 nMol) in in vitro studies using EGCG (epigallocatechin gallate). EGCG is a major catechin found in green and black tea (44). The dose-response inhibition of ENOX2 and the growth of cancer cells are well correlated (44). ENOX2 is an ongoing research target in efforts to identify the precise mechanisms of cancer cell-specific, catechin-induced growth inhibition.

Inhibition of ENOX2 by catechin-vanilloid mixtures is approximately ten to one hundred times more effective than either catechins or vanilloids acting alone. A catechin-vanilloid mixture of green tea extract equivalent to sixteen cups of green tea can significantly inhibit ENOX2 growth of cultured cancer cells.

EGCG and capsaicin have been shown to reduce the anchorage-dependent growth of HeLa (human cervical carcinoma) cells that naturally overexpress ENOX. Researchers in one study concluded that cell-surface expression of tNOX is both necessary and sufficient for the cell-surface anticancer activities of EGCG and capsaicin (44).

Another study found the combination of catechin and capsicum to be promising in inhibiting cancer cell growth. This combination may be sufficient to induce apoptosis in early-stage cancer cells prior to development of clinical symptoms when only a small number of cells are present. Research demonstrates that inhibition of ENOX2 prevents proliferation of the cells before they reach a point where they can divide (38, 39, 44, 45, 47). Cell division stops and, after several days, apoptosis ensues (38, 39, 44).

Factors Involved in the Metabolic Pathways Necessary for Cancer Cell Growth

A disruption in the primary mitochondrial pathways can result from consumption of excess calories, hypoxia, a sedentary lifestyle, environmental toxins, or genetics, all of which may affect the mitochondrial pathways and result in the following outcomes:

- excess ROS in the mitochondria (from damaged mtDNA)
- a shift from OXIPHOS ATP production to glycolysis
- downregulation of PMET (which is pertinent to cell survival because it maintains various homeostatic relationships, allowing the cell to focus on unregulated growth and cell division)
- an opening of the NOX mediated-gateway for NAD+ NADH, which allows for exponential growth of the differentiated cells

Cancer Considerations

We've framed cancer as a metabolic disease—a disease dependent on energy. As such, we're squarely focused on mitochondria and what can be done to improve the efficiency and output of the ETC.

Cancer cells need glucose for rapid growth. When glucose is restricted, cancer cells are forced to compete with healthy cells for any available glucose. While healthy cells effortlessly transition to burning ketone bodies, cancer cells are put under tremendous metabolic and oxidative pressure to use the mitochondrial OXIPHOS energy production pathway.

Throughout this book, a foundational approach of breathing, mild exercise, and a mild calorie-restricted ketogenic diet has been stressed to improve mitochondrial functionality. Whether a person is dealing with cancer or just wants to improve their overall health, a modified, calorie-restricted ketogenic diet (RKD) is an important consideration. There are many sources—both online and available from organizations—that provide help in initiating this diet. And, of course, anyone who wants to try the RKD should only do so under the supervision of a health care provider. You must be determined to be in good enough health to undergo a fast. In this diet, you consume approximately 1,400 calories/day based on a high-fat diet containing minimal sugars and carbs. Generally, this diet begins with a one- to three-day water fast. This uses up much of your glucose reserves and forces the body into using ketones for energy. If a fast seems too extreme for you, you can move into the maintenance phase of the diet by restricting carbohydrates and sugars, eating a high-fat diet, and keeping your calories to 1,400 per day.

Depending on one's health status, consideration may be also given to improving overall mitochondrial energy-conversion efficiency. This is achieved through basic mitochondrial supplement protocol:

- L-carnitine
- D-ribose
- Coenzyme Q10
- magnesium

This mitochondria protocol comprises essential supplements to help transport fats and ketone bodies into mitochondria along with inner-membrane support for the electron transport chain. If you've experienced a recent illness, consider increasing serving sizes to help improve the functionality of the mitochondria.

Oxygen and exercise are fundamental for improving the mitochondrial ETC. Low cellular oxygen content (hypoxia) drives ROS, while exercise induces production of new mitochondria. Sitting at home or in your office engaged in shallow breathing is detrimental to mitochondrial health.

This protocol can be undertaken in conjunction with conventional cancer treatments or as a standalone approach to your overall health. It has minimal side effects other than what your taste buds and appetite have become accustomed too.

Cancer is a complex illness. However, the one constant is the demand for energy to fuel the cancer cell growth. Anything that can be done to interrupt the energy supply and slow growth is the primary consideration.

References:

1. Razungles, J., Cavaillès, V., Jalaguier, S., and Teyssier, C. "The Warburg Effect: From Theory to Therapeutics in Cancer." *Médecine/Sciences* 29, no. 11 (November 2013): 1026–33.
2. Shtern, F. "Positron Emission Tomography as a Diagnostic Tool. A Reassessment Based on Literature Review." *Investigative Radiology* 27, no. 2 (February 1992): 165–8. Review.
3. Dang, C.V. "Links between Metabolism and Cancer." Genes and Development 26, no. 9 (May 1, 2012): 877–90.
4. Liu, K.C., Shih, T.Y., Kuo, C.L., Ma, Y.S., et al. "Sulforaphane Induces Cell Death through G2/M Arrest and Triggers Apoptosis in HCT 116 Human Colon Cancer Cells." *The American Journal of Chinese Medicine* 44, no. 6 (2016): 1289–1310.
5. Ganai, S.A., Rashid, R., Abdullah, E., and Altaf, M. "Plant-Derived Sulforaphane in Combinatorial Therapy against Therapeutically Challenging Pancreatic Cancer." *Anti-Cancer Agents in Medicinal Chemistry* 17, no. 3 (2017): 365–73
6. Ganai, S.A. "Histone Deacetylase Inhibitor Sulforaphane: The Phytochemical with Vibrant Activity against Prostate Cancer." *Biomedicine and Pharmacotherapy* 81 (July 2016): 250–7.
7. Jo, G.H., Kim, G., Kim, W., Park, K.Y., and Choi, Y.H. "Sulforaphane Induces Apoptosis in T24 Human Urinary Bladder Cancer Cells through a Reactive Oxygen Species-Mediated Mitochondrial Pathway: The Involvement of Endoplasmic Reticulum Stress and the Nrf2 Signaling Pathway." *International Journal of Oncology* 45, no. 4 (October 2014): 1497–1506.
8. Roy, S.K., Srivastava, R.K., and Shankar, S. "Inhibition of PI3K/AKT and MAPK/ERK Pathways Causes

Activation of FOXO Transcription Factor, Leading to Cell Cycle Arrest and Apoptosis in Pancreatic Cancer." *Journal of Molecular Signaling* 5, no. 10 (2010).

9. Bryant, C.S., Kumar, S., Chamala, S., Shah, J., et al. "Sulforaphane Induces Cell Cycle Arrest by Protecting RB-E2F-1 Complex in Epithelial Ovarian Cancer Cells." *Molecular Cancer* 9, no. 47 (2010).
10. Seyfreid, T. *Cancer Is a Metabolic Disease: On the Origin, Management, and Prevention of Cancer*, 1st ed. (Hoboken, NJ: Wiley & Sons, 2012).
11. Seyfried, T.N. and Shelton, L.M. "Cancer as a Metabolic Disease." *Nutrition and Metabolism* 7, no. 7 (Jan 27 2010).
12. Seyfried, T.N., Flores, R.E., Poff, A.M., and D'Agostino, D.P. "Cancer as a Metabolic Disease: Implications for Novel Therapeutics." *Carcinogenesis* 35, no. 3 (March 2014): 515–27.
13. Sonnenschein, C. and Soto, A.M. *The Society of Cells: Cancer and the Control of Cell Proliferation* (New York: Springer-Verlag, 1999).
14. Nixon, D. "Nutrition and Cancer: American Cancer Society Guidelines, Programs, and Initiatives." Wiley Online Library (December 2008).
15. Withrow, C.D. "The Ketogenic Diet: Mechanism of Anticonvulsant Action." *Advances in Neurology* 27 (1980): 635–42. Review.
16. Sivitz, W. and Yorek, M.A. "Mitochondrial Dysfunction in Diabetes: From Molecular Mechanisms to Functional Significance and Therapeutic Opportunities." *Antioxidants and Redox Signaling* 12, no. 4 (April 2010): 537–77.
17. Racette, S.B., Weiss, E.P., Villareal, D.T., et al. "One Year of Caloric Restriction in Humans: Feasibility and Effects on Body Composition and Abdominal Adipose Tissue." *The Journals of Gerontology, Series A,*

Biological Sciences and Medical Sciences 61, no. 9 (2006): 943–50.

18. Boland, M.L., Chourasia, A.H., and Macleod, K.F. "Mitochondrial Dysfunction in Cancer." *Frontiers in Oncology* 3 (December 2, 2013): 292.

19. Schell, J.C. and Rutter, J. "The Long and Winding Road to the Mitochondrial Pyruvate Carrier." *Cancer and Metabolism* 1 (2013): 6.

20. Brightman, A.O., Wang, J., Miu, R.K., Sun, I.L., Barr, R., Crane, F.L., and Morré, D.L. "A Growth Factor- and Hormone-Stimulated NADH Oxidase from Rat Liver Plasma Membrane." *Biochimica et Biophysica Acta* 1105 (1992): 109–17

21. Tappel, A.L. "Free-Radical Lipid Peroxidation Damage and Its Inhibition by Vitamin E and Selenium." *Federation Proceedings* 24 (January-February 1965): 73–8.

22. Lundberg, W.O. "Lipids of Biological Importance: Peroxidation Products and Inclusion Compounds of Lipids." *The American Journal of Clinical Nutrition* 6, no. 6 (November-December 1958): 601–3.

23. Sorce, S. and Krause, K.H. "NOX Enzymes in the Central Nervous System: From Signaling to Disease." *Antioxidants and Redox Signaling* 11, no. 10 (October 2009): 2481–504.

24. Leopold, J.A. and Loscalzo, J. "Oxidative Mechanisms and Atherothrombotic Cardiovascular Disease." *Drug Discovery Today: Therapeutic Strategies* 5, no. 1 (March 2008): 5–13.

25. Nayemia, Z., Jaquet, V., and Krauser, K.H. "New Insights on NOX Enzymes in the Central Nervous System." *Antioxidants and Redox Signaling* 20, no. 17 (June 10, 2014): 2815–37. *Revista Medico-Chirurgicala a Societatii De Medici Si Naturalisti Dub Iasi* 120, no. 1 (January-March 2016): 29–33. Review.

26. Clark, R., Epperson, T.K., and Valente, A.J. "Mechanisms of Activation of NADPH Oxidases." *Japanese Journal of Infectious Disease* 57, no. 5 (October 2004): S22–3. Review.

27. Baserga, R. "Oncogenes and the Strategy of Growth Factors." *Cell* 79, no. 6 (December 16, 1994): 927–30. Review.

28. Morré, D.J. and Morré, D.M. *ECTO-NOX Proteins: Growth, Cancer, and Aging* (New York: Springer, 2012), Kindle.

29. Bruno, M., Brightman, A.O., Lawrence, J., Werderitsh, D., and Morré, D.J. "Stimulation of NADH Oxidase Activity by Growth Factors and Hormones Is Decreased or Absent with Hepatoma Plasma Membranes." *Biochemical Journal* 284 (1992): 625–8.

30. Morré, D.J. and Morré, D.M. "Cell Surface NADH Oxidases (ECTO-NOX Proteins) with Roles in Cancer, Cellular Time-Keeping, Growth, Aging, and Neurodegenerative Diseases." *Free Radical Research* 37, no. 8 (August 2003): 795–808.

31. Morré, D.M., Guo, F., and Morré, D.J. "An Aging-Related Cell Surface NADH Oxidase (arNOX) Generates Superoxide and Is Inhibited by Coenzyme Q." *Molecular and Cellular Biochemistry* 254, no. 1–2 (October 2004): 101–9.

32. Herst, P.M. and Berridge, M.V. "Plasma Membrane Electron Transport: A Target for Cancer Drug Development." *Current Molecular Medicine* 6, no. 8 (December 2006): 895–904. Review.

33. Morré, D.J., Wu, L.-Y., and Morré, D.M. "The Antitumor Sulfonylurea N-(4-methylphenylsulfonyl)-N′-(4-chlorophenyl) urea (LY181984) Inhibits NADH Oxidase Activity of HeLa Plasma Membranes." *Biochimica et Biophysica Acta* 1240 (1995): 11–7.

34. Morré, D.J., McClain, N., Wu, L.-Y., Kelly, G., and

Morré, D.M. "Phenoxodiol Treatment Alters the Subsequent Response of ENOX2 (tNOX) and Growth of HeLa Cells to Paclitaxel and Cisplatin." *Molecular Biotechnology* 42, no. 1 (May 2009): 100–9.

35. De Luca, T., Morré, D.M., and Morré, D.J. "Reciprocal Relationship between Cytosolic NADH and ENOX2 Inhibition Triggers Sphingolipid-Induced Apoptosis in HeLa Cells." *Journal of Cellular Biochemistry* 110, no. 6 (August 15, 2010): 1504–11.

36. Morré, D.J. and Brightman, A.O. "NADH Oxidase of Plasma Membranes." *Journal of Bioenergetics and Biomembranes* 23, no. 3 (June 1991): 469–89. Review.

37. Morré, D.J., Korty, T., Meadows, C., Ades, L.M., and Morré, D.M. "ENOX2 Target for the Anticancer Isoflavone ME-143." *Oncology Research* 22, no, 1 (2014): 1–12.

38. Mohammad, R.M., Mugbil, I., Yediou, C., and Lowe, L. "Broad Targeting of Resistance to Apoptosis in Cancer." *Seminars in Cancer Biology* 35, supplement (December 2015): S78–103.

39. Racker, E. and Spector, M. "Warburg Effect Revisited: Merger of Biochemistry and Molecular Biology." *Science* 213, no. 4505 (July 17, 1981): 303–7.

40. Bettum, I.J., Gorad, S.S., Barkovskaya, A., Pettersen, S., Moestue, S.A., Vasiliauskaite, K., et al. "Metabolic Reprograming Supports the Invasive Phenotype in Malignant Melanoma." *Cancer Letters* 366, no. 1 (September 28, 2015): 71–83.

41. Morré, D.J. and Morré, D.M. "Role in Plasma Membrane Electron Transport," in *ECTO-NOX Proteins: Growth, Cancer, and Aging* (June 2012), 65–96. SpringerLink.

42. Szallasi, A. and Blumberg, P.M. "Vanilloid (Capsaicin) Receptors and Mechanisms." *Pharmacological Reviews* 51, no. 2 (June 1999): 159–212. Review.

43. Morré, D.M. and Morré, D.J. "Catechin-Vanilloid Synergies with Potential Clinical Applications in Cancer." *Rejuvenation Research* 9, no 1 (Spring 2006): 45–55.

44. Wu, L.Y., De Luca, T., Watanabe, T., Morré, D.M., and Morré, D.J. "Metabolite Modulation of HeLa Cell Response to ENOX2 Inhibitors EGCG and Phenoxodiol." *Biochimica et Biophysica Acta* 1810, no. 8 (August 2011): 784–9.

45. Miyase, T., Sano, M., Yoshino, K., and Nonaka, K. "Antioxidants from *Lespedeza homoloba* (II)." *Phytochemistry* 52, no. 2 (September 1999): 311–9.

46. Hanau, C., Morré, D.J., and Morré, D.M. "Cancer Prevention Trial of a Synergistic Mixture of Green Tea Concentrate plus Capsicum in a Random Population of Subjects Ages 40–84." *Clinical Proteomics* 11, no. 1 (2014): 2.

Conclusion

In November 2015, Harvard researchers, using a large national health survey that took place from 1999 to 2012, found that nearly 40 percent of Americans over age sixty-five were taking at least five pharmaceutical medications. The focus in the United States hasn't been on healthy aging and real preventive medicine but on increasing health care spending and on cosmetic treatments.

It's not a question of *if* you'll age. You will. We can't stop the aging process in any species. Aging and death are part of programmed species regulation. Therefore, the focus should be on ways to slow the aging process and keep yourself healthy during this extended lifespan. In this book, we've explored an approach to healthy aging. Anti-aging isn't the point; optimal aging is. Looking good and maintaining vitality in all aspects of life follow from optimal aging and health maintenance.

In the Introduction, I mentioned a Chinese medicine treatment theorem. The treatment principle is stated as

Yi bing tong zhi	One disease, different treatments;
Tong bing yi zh	Different diseases, one treatment.

This statement means that patients with the same disease diagnosis may receive entirely different treatments if their presenting patterns are different. Conversely, patients with different disease diagnoses may receive essentially the same treatments if their presenting patterns are the same. With that treatment principle in mind, we discussed the following age-related diseases and conditions:

- cardiovascular issues
- cardiometabolic syndrome
- neurodegenerative issues
- arthritis
- aging of the skin, eyes, and muscle
- cancer

Examination of these disease condition at the cellular level has revealed that mitochondria are a driving force in the development of diseases and disorders. Of course, there can be other causative factors, such as genetics, viruses, and environmental toxins, which means the mitochondria become a secondary force in disease progression. Whether primary or secondary, what's implicit is that mitochondrial dysfunction will accelerate the aging process and disease states.

In each disease examined, it was found that lifestyle choices are primary factors in mitochondrial dysfunction. To reiterate, there are three healthy lifestyle choices that inhibit mitochondrial dysfunction:

- Breathing to prevent hypoxia. Low levels of oxygen hamper the mitochondrial metabolic pathways of ATP production and result in low ATP output.
- Moderate exercise to drive mitochondrial biogenesis.
- A reduction in calorie intake and avoiding excessive intake of sugars and carbohydrates to prevent high glucose loads that result in increased nutrient flow to fuel mitochondrial ATP production and

consequent enhanced reactive oxygen species (ROS) production.

In the last decade, abundant research has shown that the amount of sugar consumed in the diet as well as an excess of calories are perhaps the most important factors in accelerated aging. The excess processed sugars in the American diet, especially in low-income populations, creates a glucose-signaling cellular pathway correlated with an increase in ROS or oxidant stress.

Many Americans are now taking personal responsibility for their health, educating themselves, and seeking help by integrating complementary medicine into their health care. This strategy won't only benefit the individual but will help the nation. Just the simple act of helping people to incorporate the three lifestyle changes mentioned above into their lives would, in the very short term, substantially reduce our health care costs. The first step in adopting an individualized protocol of anti-aging—or, rather, healthy longevity—is to become educated about the science behind these lifestyle choices and incorporate these strategies into your life.

Individuals who are concerned with healthy aging should now understand two fundamental concepts: One, aging isn't about superficial appearance, but one's appearance does mirror their underlying cellular health as they age; and, two, modern medicine isn't infallible in its understanding and treatment of aging. As individuals look at their own health maintenance as they age, they must make choices concerning how best to promote healthy skin, hair, muscles, joints, organs, and, most importantly, the brain.

Our journey in understanding the importance of mitochondria started with the profound significance of how life as we know it emerged on this planet, including the evolution of a particular eukaryotic cell containing an internal power source—the mitochondria. From there, we discussed the amazing self-assembling molecular machines, of which the mitochondria is one, driven by the internal thermal storm and powered by the ATP energy currency. Life happens on many levels, from colliding molecules to the symbiotic relationship of the mitochondria with the cell. All are part of who we are.

To understand the world as a whole, we need to abandon our linear, deterministic thinking. The complexity of life, our minds, and human society is played out on a network of elaborate relationships—starting with the molecules that assemble each of us. And yet, at the same time, we're very different, and each of us is unique.

The Sinatra Solution, mentioned in the Introduction to this book, impressed me because a cardiologist had written it and taken such a radical departure from his training. What clearly came through was Sinatra's curiosity (scientific inquiry) to explore and challenge traditionally held beliefs in his field.

At the end of his book, Sinatra discusses Earthing, and I thought this would be a fitting end to my book. In its simplest form, Earthing involves going outside and placing your bare feet on the ground. Wearing shoes with soles insulates you from the Earth's electromagnetic field, but, with your bare feet planted on the ground, the great waves of pulsating energy emanating from our planet freely flow through you. Most of us haven't experienced this since we were a child running on the beach or playing in our yard.

However, it's not just the energy of our planet that we feel. Our planet encircles a great electromagnetic generator—the sun—which, in turn, is part of a galaxy that's part of a star cluster and so on. Everything is connected, and just the simple act of planting your feet on the Earth connects you to the Infinite.

Maybe there really is a free, creative spirit flowing through all nature, including the dark energy or quintessence through which the cosmos is growing. Our breath is part of this universal flow. We've mechanized this flow of energy through windmills, waterwheels, steam engines, motors, and electric circuits, but, outside of man-made machines, the flow is freer. Maybe the energy balances in galaxies, stars, planets, animals, and plants aren't always exact. Energy may not always be exactly conserved.

New matter and energy may arise from quintessence, more at some times and in some places than others. The flow of energy through living organisms may not depend only on the caloric content of food and the physiology of digestion and respiration. It may also depend on the way an organism is linked to a larger flow of energy in all of nature (1).

We've explored your cellular mitochondria and the energy that drives your body at the molecular level. Where once you may have viewed yourself as solid, it's my hope in writing this book that you're left questioning this solidity. The only thing that can be truly pointed to is the existence of a consciousness—where that resides is in question. It's ultimately the Great Heart Sutra resonating through all eternity.

The Great Heart Sutra

"Listen, Sariputra,
all phenomena bear the mark of Emptiness;
their true nature is the nature of
no Birth no Death,
no Being no Non-being,
no Defilement no Purity,
no Increasing no Decreasing.
That is why, in Emptiness,
Body, Feelings, Perceptions,
Mental Formations and Consciousness
are not separate self-entities."

Reference:

1. Sheldrake, R. *Science Set Free: 10 Paths to New Discovery* (New York, NY: Crown Publishing Group, 2012), Kindle, 82.

Appendix 1

Suggested Nutraceutical Supplements

Throughout this book, research has been provided on nutritional supplements and herbal extracts that have been shown to have a positive effect on mitochondria and therefore enhance your health and lifespan. This section provides links to some of the nutritional supplements discussed previously.

I want to emphasize that the basic mitochondria protocol must be coupled with the following practices:

- Breathing to prevent hypoxia. Low levels of oxygen hamper the mitochondrial metabolic pathways of ATP production and result in low ATP output.
- Moderate exercise to drive mitochondrial biogenesis.
- A reduction in calorie intake and avoiding excessive intake of sugars and carbohydrates to avoid high glucose loads. This results in increased nutrient flow to mitochondria for ATP production and consequent enhanced ROS production.

Basic Mitochondria Protocol

The basic protocol is specifically used to support movement of fats into the mitochondria (the preferred energy source), supporting the rebuilding of inner mitochondrial membranes damaged by free radicals, and improving the electron transport chain (ETC) for ATP production.

The basic mitochondria protocol is available as a kit or individual nutritional supplements:

Basic Mito Kit
Two servings twice/day
https://www.themitobook.com/mitochondria-basic-kit-products-501_508.html

D-ribose
Serving = 1 g/day
https://www.themitobook.com/now-d-ribose-powder-8-ounce-p-70935.html

L-Carnitine
Serving = 2 g/day
https://www.themitobook.com/now-l-carnitine-liquid-16-ounce-3000-milligrams-citrus-flavor-p-70934.html

Coenzyme Q10 (Ubiquinol)
Serving = 100 mg, twice/day
https://www.themitobook.com/tishcon-corp.-quinogel-solubilized-ubiquinol-coq10-(hydrosoluble-kaneka-qh)-60-soft-gels-100-milligrams-p-69492.html

Magnesium
Serving = 100 mg, twice/day
https://www.themitobook.com/now-magnesium-citrate-120-veggie-capsules-p-70936.html

Rationale:
D-ribose—promotes energy transformation
D-ribose is a simple, five-carbon sugar that's an intermediate in the pentose phosphate pathway that produces genetic materials and provides substrates for fatty acids and hormones. D-ribose restores the purine sources lost when ADP+ADP combine to make ATP+AMP (AMP must be discarded).

Coenzyme Q10: The Most Important Nutrient for

Mitochondrial Health!
CoQ10 is naturally occurring in almost every cell of the body and is an essential component of life. As we age, we produce less and less CoQ10. CoQ10 is an antioxidant and a cofactor in the mitochondrial ETC. Free radicals escape from the mitochondrial ETC, and CoQ10 is able to act as an antioxidant and pick up these free-radical oxygen species and return them to the ETC. Thus, CoQ10 protects mtDNA, lipid membrane structure, and other proteins from free-radical damage.

L-Carnitine
Anaerobic metabolism produces lactate, which can promote angina. Carnitine helps to facilitate the aerobic processes, delay a shift to anaerobic metabolism, and prevent the production of lactate because carnitine is the sole transporter of fats and triglycerides into mitochondrial Complexes II through V.

L-carnitine is a naturally occurring compound synthesized in our bodies from lysine and methionine, but stores decrease as we grow older. L-carnitine is primarily a transporter of long-chain fatty acids into the mitochondria (Complexes II through V), where they undergo beta-oxidation to produce ATP (fatty acids can't enter the mitochondria on their own; they must attach an acyl group to the chain). Mitochondria prefer fatty acids for fuel. Almost all dietary fatty acids are long-chain. The rates of oxidative phosphorylation, beta-oxidation, and energy turnover are contingent on L-carnitine availability.

Dietary L-carnitine is found mostly in animal products, but vegetarians aren't usually deficient because it can be synthesized in the body.

Uses: Aside from fatty-acid transfer, L-carnitine also reduces lactic acid buildup from anaerobic metabolism. It removes excess acyl groups residing in the mitochondria, which can disturb the metabolism of fats. Lactic acid caused tissue and muscle damage. L-carnitine also supports cognitive function in Alzheimer's disease and other dementias, improves memory, and reduces fatigue.

Magnesium

Magnesium is involved in more than three hundred biochemical reactions in the body, including the production of ATP. Uses: Magnesium plays a role in muscle relaxation by assisting ATP and enzymes to relax the heart muscle. Calcium initiates muscle contraction and, without magnesium, calcium remains in the muscle cell and the heart remains contracted. Contraction results in vasoconstriction, which slows oxygen delivery and inhibits mitochondrial energy production. Magnesium deficiency is tied to numerous conditions, including diabetes, ischemic heart disease, congestive heart failure, angina, asthma, insulin resistance, cardiomyopathies, etc.

Cardiovascular Support

For purposes of enhancing cardiac function, the following supplements would be added to the basic mitochondria protocol:

Basic Mito Kit

Two servings twice/day

https://www.themitobook.com/basic-mito-kit.html

Alpha Lipoic Acid

https://www.themitobook.com/geronova-research-r-lipoic-acid-stabilized-bio-enhanced-60-veggie-capsules-300-milligrams-p-70717.html

Nicotinamide Riboside

https://www.themitobook.com/thorne-research-niacel-250-60-count-p-69652.html

Alpha Lipoic Acid (ALA)
ALA is a naturally occurring antioxidant found in the mitochondria, making it very rare.
ALA benefits arise from its remarkable ability to dramatically improve glucose control and restore insulin sensitivity.

Uses: ALA regulates NAD and NADH levels in the mitochondria. With high glucose levels, cells can't convert NADH to NAD. The cell can't access NAD, and excess NADH causes free-radical damage, a breakdown of iron storages, and a backup of electrons at Complex I in the ETC. Additionally, ALA has been used as an anti-aging compound for restoration of cognitive function, heart function, and activity levels.

Nicotinamide Riboside
Clinical research demonstrates that nicotinamide riboside (NR) is a highly efficient precursor to NAD+. Incorporating NR into the body has been shown to restore NAD+ levels in laboratory models, combating the age-related decline of the NAD+ supply in cells. Preclinical studies link the restoration of NAD+ supplies to an improvement in metabolic efficiency and overall health in aging mitochondria.

Neurocognitive Support
To enhance neurological and cognitive function, the following supplements would be added to the basic mitochondria protocol:

Basic Mito Kit
Two servings twice/day
https://www.themitobook.com/mitochondria-basic-kit-products-501_508.html

Alpha Lipoic Acid
https://www.themitobook.com/geronova-research-r-lipoic-acid-stabilized-bio-enhanced-60-veggie-capsules-300-

milligrams-p-70717.html

Curcumin
https://www.themitobook.com/qol-labs-unisorb-curcumin-30-veggie-capsules-p-70678.html

EGCG
https://www.themitobook.com/biotics-research-egcg-200mg-60-capsules-p-69092.html

Resveratrol
https://www.themitobook.com/designs-for-health-resveratrol-supreme-60-capsules-p-55839.html

Rationale:

Alpha-Lipoic Acid (ALA)

ALA is a naturally occurring antioxidant found in the mitochondria, making it very rare.
ALA benefits arise from its remarkable ability to dramatically improve glucose control and restore insulin sensitivity.

Uses: ALA regulates NAD and NADH levels in the mitochondria. With high glucose levels, cells can't convert NADH to NAD. The cell can't access NAD, and excess NADH causes free-radical damage, a breakdown of iron storages, and a backup of electrons at Complex I in the ETC. Additionally, ALA has been used as an anti-aging compound for restoration of cognitive function, heart function, and activity levels.

Curcumin

Curcumin has been reported to cross the blood-brain barrier and has an outstanding safety profile with potential for neuroprotective efficacy, including anti-inflammatory, antioxidant, and anti-protein-aggregate activities. As an antioxidant, curcumin promotes and supports brain health by fighting free radicals, which can cause oxidative stress for cells. This antioxidant protection has wide-ranging effects, from helping sustain memory function to promoting optimal cognitive function

EGCG

EGCG has been shown to reduce both amyloid production in cultured cells and Aβ-amyloid plaque deposition in transgenic mice. A study aimed at investigating the association between green tea consumption and cognitive function in elderly Japanese subjects showed that a higher consumption of this beverage was associated with a lower prevalence of cognitive impairment.

Resveratrol

Resveratrol is a natural polyphenolic compound mainly found in the skin of grapes and is well known for its phytoestrogenic and antioxidant properties. Resveratrol helps impaired mitochondria switch from fatty-acid storage to oxidation. Fatty-acid storage contributes considerably to intramyocellular lipid accumulation, which has been linked to insulin resistance in obese individuals and type 2 diabetes. Resveratrol improves insulin resistance, protects against diet-induced obesity, induces genes for oxidative phosphorylation, and activates PGC-1α.

Blood Sugar Management

For management of blood glucose levels, the following supplements would be added to the basic mitochondria protocol:

Basic Mito Kit

Two servings twice/day

https://www.themitobook.com/basic-mito-kit.html

Alpha Lipoic Acid

https://www.themitobook.com/geronova-research-r-lipoic-acid-stabilized-bio-enhanced-60-veggie-capsules-300-

milligrams-p-70717.html

Resveratrol
https://www.themitobook.com/designs-for-health-resveratrol-supreme-60-capsules-p-55839.html

Shilajit
https://www.themitobook.com/mother-earth-labs-fulvic-liquid-minerals-4-oz-super-concentrate-400x-p-67314.html?

Nicotinamide Riboside
https://www.themitobook.com/thorne-research-niacel-250-60-count-p-69652.html

Rationale:

Alpha-Lipoic Acid (ALA)

ALA is a naturally occurring antioxidant found in the mitochondria, making it very rare.

ALA benefits arise from its remarkable ability to dramatically improve glucose control and restore insulin sensitivity.

Uses: ALA regulates NAD and NADH levels in the mitochondria. With high glucose levels, cells can't convert NADH to NAD. The cell can't access NAD, and excess NADH causes free-radical damage, a breakdown of iron storages, and a backup of electrons at Complex I in the ETC. Additionally, ALA has been used as an anti-aging compound for restoration of cognitive function, heart function, and activity levels.

Resveratrol

Resveratrol is a natural polyphenolic compound mainly found in the skin of grapes and is well known for its phytoestrogenic and antioxidant properties. Resveratrol helps impaired mitochondria switch from fatty-acid storage to oxidation. Fatty-acid storage contributes considerably to intramyocellular lipid accumulation, which has been linked to insulin resistance in obese individuals and type 2 diabetes. Resveratrol improves insulin resistance, protects against diet-induced obesity, induces genes for oxidative phosphorylation, and activates PGC-1α.

Shilajit

Researchers have shown that a nutrient-rich biomass called shilajit (fulvic acid) can boost CoQ10 efficiency. Shilajit has beneficial effects on cellular energy, diabetes, and memory, and it protects against cognitive decline. Shilajit helps combat mitochondrial dysfunction-induced aging. Working synergistically with CoQ10, shilajit boosts energy, protects mitochondria, and reduces aging at the cellular level.

Reducing the Effects of Aging in the Skin and Eyes

To protect skin and eyes from the effects of aging, the following supplements would be added to the Mitochondria Basic Protocol:

Basic Mito Kit

Two servings twice/day

https://www.themitobook.com/mitochondria-basic-kit-products-501_508.html

L-Carnosine

https://www.themitobook.com/pure-encapsulations-l-carnosine-60-vegetarian-capsules-500-milligrams-p-44727.html

Resveratrol

https://www.themitobook.com/designs-for-health-resveratrol-supreme-60-capsules-p-55839.html

Shilajit

https://www.themitobook.com/mother-earth-labs-fulvic-liquid-minerals-4-oz-super-concentrate-400x-p-67314.html

Nicotinamide Riboside
https://www.themitobook.com/thorne-research-niacel-250-60-count-p-69652.html

L-Carnosine
L-carnosine is a combination of two amino acids—beta-alanine and histidine. Research has found carnosine to be a potent inhibitor of glycation (remember, glycation is the cross-linking in cells that causes wrinkles in addition to other unhealthy effects). In in vitro studies, carnosine has been able to reverse damage to cells caused by aging, returning them to more youthful and efficient function. In one human study, fifteen healthy subjects were given histidine and carnosine. The researchers found that these compounds protected the LDL (bad cholesterol) from glycation and oxidation.

Resveratrol
Resveratrol is a natural polyphenolic compound mainly found in the skin of grapes and is well known for its phytoestrogenic and antioxidant properties. Resveratrol helps impaired mitochondria switch from fatty-acid storage to oxidation. Fatty-acid storage contributes considerably to intramyocellular lipid accumulation, which has been linked to insulin resistance in obese individuals and type 2 diabetes. Resveratrol improves insulin resistance, protects against diet-induced obesity, induces genes for oxidative phosphorylation, and activates PGC-1α.

Shilajit
Researchers have shown that a nutrient-rich biomass called shilajit (fulvic acid) can boost CoQ10 efficiency. Shilajit has beneficial effects on cellular energy, diabetes, and memory, and it protects against cognitive decline. Shilajit helps combat

mitochondrial dysfunction-induced aging. Working synergistically with CoQ10, shilajit boosts energy, protects mitochondria, and reduces aging at the cellular level.

Nicotinamide Riboside

Clinical research demonstrates that nicotinamide riboside (NR) is a highly efficient precursor to NAD+. Incorporating NR into the body has been shown to restore NAD+ levels in laboratory models, combating the age-related decline of the NAD+ supply in cells. Preclinical studies link the restoration of NAD+ supplies to an improvement in metabolic efficiency and overall health in aging mitochondria.

Joint Support

To protect against degenerative damage to the joints—in particular, arthritis—the following supplements would be added to the basic mitochondria protocol:

Basic Mito Kit

Two servings twice/day

https://www.themitobook.com/mitochondria-basic-kit-products-501_508.html

Resveratrol

https://www.themitobook.com/designs-for-health-resveratrol-supreme-60-capsules-p-55839.html

Alpha Lipoic Acid

https://www.themitobook.com/geronova-research-r-lipoic-acid-stabilized-bio-enhanced-60-veggie-capsules-300-milligrams-p-70717.html

Methylsulfonylmethane

https://www.themitobook.com/kala-health-msm-powder-1-pound-coarse-flakes-p-70695.html

Hydroxytyrosol
https://www.themitobook.com/joy-of-health-olea25-30-capsules-100-milligrams-p-70681.html

NT Factor
https://www.acuatlanta.net/allergy-research-group-nt-factors-advanced-physician-formula-150-count-p-61490.html

Rationale:

Alpha Lipoic Acid

ALA is a naturally occurring antioxidant found in the mitochondria, making it very rare.
ALA benefits arise from its remarkable ability to dramatically improve glucose control and restore insulin sensitivity.

Uses: ALA regulates NAD and NADH levels in the mitochondria. With high glucose levels, cells can't convert NADH to NAD. The cell can't access NAD, and excess NADH causes free-radical damage, a breakdown of iron storages, and a backup of electrons at Complex I in the ETC. Additionally, ALA has been used as an anti-aging compound for restoration of cognitive function, heart function, and activity levels.

Resveratrol

Resveratrol is a natural polyphenolic compound mainly found in the skin of grapes and is well known for its phytoestrogenic and antioxidant properties. Resveratrol helps impaired mitochondria switch from fatty-acid storage to oxidation. Fatty-acid storage contributes considerably to intramyocellular lipid accumulation, which has been linked to insulin resistance in obese individuals and people with type 2 diabetes. Resveratrol improves insulin resistance, protects against diet-induced obesity, induces genes for oxidative phosphorylation, and activates PGC-1α.

The most important properties of resveratrol are its beneficial effects on oxidative stress, vascular inflammation, and platelet aggregation. In fact, resveratrol upregulates the endogenous antioxidant systems, such as superoxide dismutase (SOD) enzymes.

Methylsulfonylmethane (MSM)

Methylsulfonylmethane (MSM) provides biologically active sulfur, which is the fourth-most plentiful mineral in the body and needed for many different critical bodily functions. A well-researched benefit of MSM is that it helps decrease joint inflammation, improves flexibility, and enhances collagen production. MSM supplements are beneficial for helping the body form new joint and muscle tissue while lowering inflammatory responses that contribute to swelling and stiffness. An MSM supplement serves as a natural and effective anti-inflammatory because sulfur stimulates the immune system and facilitates normal cellular activity. Sulfur needs to be present for our cells to release many byproducts and excess fluids that can accumulate in tissues and cause swelling/tenderness.

Hydroxytyrosol

Hydroxytyrosol is an anti-inflammatory found in extra-virgin olive oil, but it has benefits beyond inhibiting inflammation. Several studies have shown benefits for heart health, protecting against bone loss, and preventing the development of neurological diseases. A study published in the journal *Molecules* in 2014 discussed the protective effects of hydroxytyrosol on the neurological system. In an animal study published in the peer-reviewed journal *PLOS ONE* in 2014, researchers showed that, when virgin olive oil was combined with vitamin D and used as a supplement, it protected against bone loss.

NT Factor™

NT FactorTM formulation is specially designed so the

phosphoglycolipids it contains match those in the human cell membrane so they're not digested but instead are absorbed by the walls of the gastrointestinal tract intact. Once there, the phosphoglycolipids go to work repairing any damage in the cell membranes. As we get older or develop various illnesses, oxidative stress produces free radicals that destroy cell membranes, literally creating holes in our cells. The cell membrane is composed mainly of fats and some protein. When the phosphoglycolipids in NT Factor get into the cell membrane, they fill in and plug any holes. This enables the cell to once again function as it's meant to.

Appendix 2

All Calories Aren't Created Equal

The terms "caloric excess" and "calorie restriction" are used throughout this book. Calories are the energy source that fuels mitochondria. The human body requires energy in order to operate. Everything from brain activity to blood flow requires energy, which is where calories come in.

Conventional thinking assumes that the interaction between food and the human body is as follows:

Calories in minus Calories Expended = Calorie Deficit/Surplus. From this formula, we can construct an example. Let's say your body requires 2,000 calories every day to keep your physiology functioning. If you consume 1,800 calories, you'll be operating under a caloric deficit, and your body will seek the necessary extra energy from another source (such as your fat reserves). Conversely, if you consume 2,200 calories, your body will store the surplus energy as fat. In its simplest form, this formula defines the relationship between caloric excess or deficit, but it's more complex than this.

Many diets that use the simple approach of calories in – calories expended = calorie deficit/surplus are based on a calorie-counting approach. The problem with this approach is that not all sources of calories are equal in meeting the energy needs of your mitochondria. The way in which different types of foods influence the chemical reactions within our body has a huge effect on the percentage of the calories we consume that will ultimately be converted into fat.

Our diets consist of three main nutrients: protein, fat, and carbohydrates.

Protein contains 4 kcals (kilocalories) per gram (1 kcal is equal to 1,000 calories). Protein is found in animal sources, such as meat, fish, and dairy products, but it's also available in a wide variety of other sources, such as whole grains, legumes, nuts, and seeds.

Protein is the second-most abundant molecule in the body (after water). It's required for a number of functions within the human body—everything from building and repairing muscle tissue to replicating DNA. It can't be wholly synthesized by the human body and must therefore be supplied, in part, through diet.

Fat contains 9 kcals per gram. "Fat" is a general term for a number of different compounds that share key characteristics. In terms of what you eat, fats are found in a wide variety of sources, such as oils, butter, and nuts.

Fat has a number of functions within the body. It's most commonly understood to be a source of energy (within fat reserves), but it's also vital for the absorption of certain vitamins, maintaining healthy skin and hair, and maintaining body temperature. Fat can be stored within the body and then converted into glucose as needed as an energy source. However, what's not widely known is that fats are the preferred energy source for your mitochondria.

Carbohydrates contain 4 kcals per gram. Highly concentrated carbohydrates are found in a wide range of refined foods, such as bread, pasta, breakfast cereals, candy, and chocolate. Carbs also predominate in unrefined foods, such as beans, tubers, and rice.

Carbohydrates are typically broken down into glucose to be used as energy in the body. They're what your body will call upon first for functions beyond those that require protein or fat, yet carbohydrates are completely unnecessary for life. That's right. You can exist just fine without them. If you consume no carbohydrates, your body will synthesize the glucose it needs for energy from available protein and/or fat in the body.

At this point, you understand that a calorie of protein isn't the same as a calorie of fat or a calorie of carbohydrate. This brings us to the crux of the issue regarding caloric excess and caloric restrictions. Consider the sedentary person who consumes 2,500 to 3,000 calories a day, most of which are carbohydrates. This person is operating under a caloric excess. The excess carbs drive fat accumulation and glycolysis—the cellular production of glucose that occurs outside the mitochondria. As was discussed throughout this book, chronic use of the glycolytic energy pathway put the mitochondria in "idle," creating reactive oxygen species (ROS) that drive age-related diseases, including cancer. In fact, many cancers are now identified by the excessive amount of glucose that cancerous cells use to proliferate.

For this sedentary person, a calorie-restrictive diet would involve a 20 percent reduction in overall calories, but, most importantly, the elimination or substantial reduction of carbs in the diet. This diet would be coupled with some form of exercise to re-engage the idle mitochondria into energy production to burn excess calories.

The takeaway here is: Fat is necessary for human life and really rather good for you—especially if you stick to the essential fatty acids. You need fat. What you don't need are carbohydrates.

When it comes to the need for carbs, our bodies are living in a past when food was scarce and excess fat stores were a good thing. Our bodies didn't know if there would be as many carbohydrates available to them tomorrow as today, so they grabbed what carbs they could when they found them. At least in developed countries, finding sufficient food and carbs isn't usually a problem.

What It Means to Be Healthy Enough

It makes logical sense that a diet high in the nutrients the body needs most would be good for you. But what does this mean for us? Should we all immediately jump on high-protein, low-carbohydrate diets? With the above in mind, let's look at a three simple adjustments you can make to what you eat that will enable you to nourish your mitochondria, have more energy, and lose weight.

1. Eat a Protein-Rich Breakfast

Replace your cereal, bagel, or toast with bacon and eggs. Go for free-range or organic if you can afford it to derive the best nutritional value, but don't sweat it if you can't.

2. Replace Carbs

Replace your rice and potatoes with steamed vegetables dressed with good-quality butter or olive oil. You'll still be eating the same amount of food (i.e., your eyes will see a full plate and tell your brain you're not starving yourself), but the carb hit will be much lower.

3. Swap the Carbs-to-Protein Ratio in Your Meals

If you can't bear to let go of your favorite carbs, try swapping your carbs-to-protein ratio. Most meals have a high carb-to-protein ratio—i.e., you'll have a load of carbs on your plate and a relatively low portion of protein. Turn that ratio on its head. Have twice as much meatloaf and a small amount of potatoes. You can achieve this with just about any meal, and you probably won't even notice the difference. Even if you do, you probably won't mind.

Put simply, if you adopt a low-carb diet, you'll almost certainly not gain weight (regardless of how much you eat), consume enough food to feel satisfied, and keep hunger at bay while maintaining or losing weight. This is definitely the "healthy enough" way. Understanding the effect of different nutrients on mitochondrial health and the development of age-related diseases should be enough to discourage you from gorging on carbohydrates.

GLOSSARY

Adenosine: a compound formed by combining a purine ring (adenine) with D-ribose.

ADP: adenosine diphosphate, the precursor of ATP.

AMP: adenosine monophosphate; a byproduct formed when two ADP molecules combine to form ATP.

Anaerobic metabolism: energy production in the cell (mostly the cytosol) that doesn't require oxygen; important in providing short bursts of energy quickly but an inefficient use of fuel that can't sustain the cell long term.

Antioxidant: any compound that protects against oxidation (free-radical damage) either by directly sacrificing itself (to protect other molecules) or indirectly by catalyzing the breakdown of biological oxidants.

Apoptosis: programmed cell death, or cellular suicide; a finely coordinated and carefully controlled mechanism for removing damaged or unnecessary cells from a multicellular organism.

ATP: adenosine triphosphate; the universal energy currency of life formed from ADP (adenosine diphosphate) and phosphate; splitting ATP releases energy used to power many different types of biochemical work, from muscular contraction to protein synthesis.

ATPase: also known as ATP synthase, this enzymatic motor is imbedded within the inner mitochondrial membrane and forms ATP (from ADP and phosphate) as protons flow through it.

Cardiomyocyte: a heart muscle cell (myocyte=muscle cell)

Cell: the smallest biological unit capable of independent life, which is able to carry out the functions of self-replication and metabolism.

Cell membrane: the permeable outer "shell" of our cells; maintains the integrity of the internal cell environment; allows nutrients to pass in and waste to pass out.

Chromosome: the long molecule of DNA; may be circular, as in bacteria and mitochondria, or straight, as in the nucleus of eukaryotic cells (in which it's wrapped in proteins such as histones).

Clonal expansion: the production of daughter cells all originally arising from a single cell.

Cross-linking of proteins by glucose: is the result of the typically covalent bonding of a sugar molecule, such as glucose or fructose, to a protein or lipid molecule.

Cytochrome c: a mitochondrial protein that shuttles electrons from Complex III to Complex IV of the electron transport chain (ETC); when released from the inner mitochondrial membrane, cytochrome c is an important initiator of apoptosis.

Cytoplasm: everything inside the cell contained within the cell membrane, excluding the nucleus.

Cytosol: the aqueous part of the cytoplasm, excluding organelles such as mitochondria and membrane systems.

DNA: deoxyribonucleic acid, the double-helix structure composed of molecules, in which nucleotide letters are paired with each other to form a template from which an exact copy of the whole molecule can be regenerated; the sequence of nucleotide letters in a gene encodes the sequence of amino acids in a protein.

ECTO-NOX: ecto-nicotinamide dinucleotide oxidase disulfide thiol exchanger; a family of cell-surface proteins on the outer membrane with CoQH2 oxidase and engaged in protein interchange activities. These proteins are often released into the extracellular fluid of the cell.

Electron: a tiny, negatively charged wave particle.

Enzyme: a protein molecule with enormous specificity that serves as a catalyst responsible for significantly speeding up biochemical reactions.

Eukaryotic cell: cells with a true nucleus.

Excitotoxicity: the process by which nerve cells are damaged by overstimulation of neurotransmitters, causing neurodegeneration and damage to cellular components.

Free radical: a highly reactive atom or molecule with an unpaired electron.

Free-radical leakage: continuous, low-level production of free radicals from the electron transport chains of the mitochondria; a result of electrons reacting directly with oxygen.

Gene: a stretch of DNA whose sequence of letters encodes for a single protein.

Genome: the complete library of genes in an organism; the term is also taken to include noncoding (i.e., nongenetic) stretches of DNA.

Glucose cross-linking: sugars bind to carbon molecules and proteins to form cross-links in the cellular matrix; this is widely implicated in extracellular matrix damage in aging and diabetes.

The cross-linking interrupts the flow of nutrients and the removal of toxins from the cells.

Glycolysis: the first step in the breakdown of glucose to extract energy for cellular metabolism. Most living organisms carry out glycolysis as part of their metabolism.

Histones: protective proteins that bind DNA in a very particular way, found mainly in eukaryotic cells.

Hypoxia: a situation that occurs when cells or tissues are deprived of oxygen.

Intermediate: any chemical substance produced during the conversion of a reactant to a product.

Intermembrane: the space between two membranes, as in a cell or organelle.

Ischemia: a situation in which there's reduced blood flow to a tissue/organ; reduced blood flow results in hypoxia.

Krebs cycle: also known as the tricarboxylic acid (TCA) cycle and/or citric acid cycle; a metabolic pathway in the mitochondria that converts carbohydrates, fats, and proteins into energy compounds (NADH and FADH2), which then enter the electron transport chain to ultimately create ATP.

Lipid: a type of long-chain, fatty molecule found in biological membranes and stored as fuel.

Membrane: the thin, fatty (lipid) layer that envelops cells and forms complex systems inside eukaryotic cells.

Metabolic rate: the rate of fuel consumption or energy production, measured by the rate of glucose oxidation or oxygen consumption.

Mitochondria: a spherical or elongated organelle in the cytoplasm of nearly all eukaryotic cells containing genetic material and enzymes important for cell metabolism, including those responsible for the conversion of food to usable energy.

Mitochondrial DNA: the chromosome found in mitochondria; five to ten copies are typically found in each mitochondrion; circular and bacterial in nature.

Mitochondrial Eve: the most recent female ancestor common to all humans living today, as determined by mitochondrial DNA inherited asexually down the maternal line.

Mitochondrial genes: genes encoded by mitochondrial DNA; in humans, there are thirteen protein-coding genes in addition to genes encoding RNA ribosomes.

Mitochondrial respiratory chain complexes: genes that provide instructions for proteins involved in oxidative phosphorylation, also called the respiratory chain. Five protein complexes, made up of several proteins each, are involved in this process. The complexes are named Complex I, Complex II, Complex III, Complex IV, and Complex V.

Multicellular: having or consisting of many cells.

Mutation: an inherited or acquired change in the DNA sequence; can have a negative, positive, or neutral effect on function.

NADH: nicotinamide adenine dinucleotide; a molecule that ultimately carries the electrons derived from food/fuel to Complex I of the electron transport chain.

Non-coding (junk) DNA: DNA sequences that don't code for proteins or RNA.

Nucleus: spherical, membrane-enclosed "control center" of eukaryotic cells; contains chromosomes composed of DNA and protein.

Organelles: tiny organs, such as mitochondria, within cells that are dedicated to specific tasks.

Oxidation: loss of electrons from an atom or molecule.

Oxidative phosphorylation: the metabolic pathway in which cells use enzymes to oxidize nutrients, thereby releasing energy that's used to re-form ATP. In most eukaryotes, this takes place inside mitochondria.

Pentose phosphate pathway: biological pathway that runs in concert with glycolysis to produce nucleotides, the building blocks of genetic materials; oxidizes glucose to form NADPH and five carbon sugars.

Phagocytosis: physical engulfment (by means of changing shape) of dead cells, pathogens, or particles by a cell; the particles are digested in a vacuole inside the cell.

PMET: plasma membrane electron transport; glycolytic cells have been shown to consume oxygen on the cell surface via plasma membrane electron transport (PMET), a process that oxidizes intracellular NADH, supports glycolytic ATP production, and may contribute to aerobic energy production.

Prokaryote: a broad class of single-celled organisms, including bacteria, that don't possess a nucleus.

Protein cross-linking: a bond, atom, or group linking the chains of atoms in a protein molecule.

Proton: a subatomic particle with a positive charge found in the nucleus of an atom.

Proton gradient: a difference in proton concentration between one side of a membrane and the other.

Purines: one of the key building blocks of DNA, RNA, and ATP; adenine is a purine.

Purine pool: an equilibrium of purine nucleotides that exists freely in the cell, commonly used in processes to construct DNA, RNA, and ATP.

Redox reaction: a reaction between two molecules in which one is oxidized (loses an electron) while the other is reduced (gains an electron).

Redox signaling: the change in activity (usually by free radicals) of transcription factors as a result of their oxidation or reduction state.

Reduction: gain of electrons by an atom or molecule.

Respiration: oxidation of food/fuel to generate energy in the form of ATP.

Respiratory chain: also known as the electron transport chain (ETC), the series of complexes embedded in the inner mitochondrial membrane that pass electrons derived from fuel from one complex to the next complex. The energy released by the transport of electrons is used to pump protons across the membrane.

Reperfusion: the restoration of oxygen into tissues suffering ischemia (blockage of blood flow) and hypoxia after stroke or heart attack.

RNA: ribonucleic acid; includes messenger RNA (an exact copy of the DNA sequence in an individual gene, contained in the cytoplasm); ribosomal RNA (forms part of the ribosomes, the protein-building factories); and transfer RNA (an adapter that couples a genetic code to a particular amino acid).

Signaling cascade: the transduction (transfer) of an initial signal through a cascade of biochemical reactions within a cell. The progression of events usually goes as follows: extracellular signaling molecule binds to a receptor protein on the outside of a cell to initiate a signaling cascade of a series of reactions with the help of several protein complexes to achieve a desired outcome.

Superoxide radical attack: two electrons are required for each oxygen atom to form H_2O; if only one electron is given to each oxygen atom, a superoxide radical is the result. Superoxide radicals tend to be quite reactive and can damage many of the most important macromolecules in the body as well as DNA strands.

Symbiosis: a mutually beneficial relationship between two organisms.

Transcription factor: a protein that binds to a DNA sequence, signaling the transcription of that gene into an RNA copy (the first step in protein synthesis).

Transfer RNA: an adaptor molecule composed of RNA, typically seventy-six to ninety nucleotides in length, that serves as the physical link between the messenger RNA (mRNA) and the amino acid sequence of proteins.

Uncoupling: disconnecting oxidative phosphorylation from ATP production; instead, the proton gradient is dissipated by protons passing back through membrane pores (instead of ATPase), creating heat.

Uncoupling protein: a channel in the membrane that allows protons to flow back through the membrane, dissipating the proton gradient as heat.

Vasoconstriction: the constriction of blood vessels; in the heart, vasoconstriction is promoted by calcium remaining in the cardiac muscles.

ABOUT THE AUTHOR

Warren L. Cargal, L. Ac.

Warren L. Cargal is a licensed acupuncturist, herbalist, and clinic director at Acupuncture Atlanta. He's practiced Chinese medicine for more than twenty years in the fields of infertility and chronic disease, and he has helped many men and women. He maintains an active clinical practice in Atlanta, Georgia.

His interest is in the integration of classical Chinese medicine with modern scientific study and evidence-based protocols. He's spent hundreds of post-graduate hours studying Chinese herbology, nutrition, and endocrine education.

Through work at his clinic, he recognized three fundamental factors that drive aging and the age-related diseases of diabetes, cardiovascular disease, neurodegenerative disease, and cancer. Those factors are shallow breathing, excessive consumption of calories and carbohydrates, and lack of exercise. This book, *Your Mitochondria: Key to Health and Longevity*, which is based on the latest research on herbal isolates and nutraceuticals, presents approaches for correcting these three factors to reverse age-related diseases.

Warren is also the creator of www.acuatlanta.net, which provides trusted and credible information on Chinese herbal formulas and nutritional supplements. The site offers access to nutritional consultants who can help you take charge of your health and lead a vibrant life filled with passion and purpose.

Warren is a student and practitioner of the Mahamudra tradition of Buddhism as a direct path to nondual awareness. He's married with one son and enjoys cross-country biking.

He can be contacted at wcargal9@acuatlanta.net.

Printed in Great Britain
by Amazon